Lecture Notes in Physics

Monographs

Editorial Board

R. Beig, Wien, Austria
J. Ehlers, Potsdam, Germany
U. Frisch, Nice, France
K. Hepp, Zürich, Switzerland
W. Hillebrandt, Garching, Germany
D. Imboden, Zürich, Switzerland
R. L. Jaffe, Cambridge, MA, USA
R. Kippenhahn, Göttingen, Germany
R. Lipowsky, Golm, Germany
H. v. Löhneysen, Karlsruhe, Germany
I. Ojima, Kyoto, Japan
H. A. Weidenmüller, Heidelberg, Germany
J. Wess, München, Germany
J. Zittartz, Köln, Germany

Managing Editor

W. Beiglböck
c/o Springer-Verlag, Physics Editorial Department II
Tiergartenstrasse 17, 69121 Heidelberg, Germany

Springer
Berlin
Heidelberg
New York
Barcelona
Hong Kong
London
Milan
Paris
Tokyo

Physics and Astronomy ONLINE LIBRARY

http://www.springer.de/phys/

The Editorial Policy for Monographs

The series Lecture Notes in Physics reports new developments in physical research and teaching - quickly, informally, and at a high level. The type of material considered for publication in the monograph Series includes monographs presenting original research or new angles in a classical field. The timeliness of a manuscript is more important than its form, which may be preliminary or tentative. Manuscripts should be reasonably self-contained. They will often present not only results of the author(s) but also related work by other people and will provide sufficient motivation, examples, and applications.
The manuscripts or a detailed description thereof should be submitted either to one of the series editors or to the managing editor. The proposal is then carefully refereed. A final decision concerning publication can often only be made on the basis of the complete manuscript, but otherwise the editors will try to make a preliminary decision as definite as they can on the basis of the available information.
Manuscripts should be no less than 100 and preferably no more than 400 pages in length. Final manuscripts should be in English. They should include a table of contents and an informative introduction accessible also to readers not particularly familiar with the topic treated. Authors are free to use the material in other publications. However, if extensive use is made elsewhere, the publisher should be informed. Authors receive jointly 30 complimentary copies of their book. They are entitled to purchase further copies of their book at a reduced rate. No reprints of individual contributions can be supplied. No royalty is paid on Lecture Notes in Physics volumes. Commitment to publish is made by letter of interest rather than by signing a formal contract. Springer-Verlag secures the copyright for each volume.

The Production Process

The books are hardbound, and quality paper appropriate to the needs of the author(s) is used. Publication time is about ten weeks. More than twenty years of experience guarantee authors the best possible service. To reach the goal of rapid publication at a low price the technique of photographic reproduction from a camera-ready manuscript was chosen. This process shifts the main responsibility for the technical quality considerably from the publisher to the author. We therefore urge all authors to observe very carefully our guidelines for the preparation of camera-ready manuscripts, which we will supply on request. This applies especially to the quality of figures and halftones submitted for publication. Figures should be submitted as originals or glossy prints, as very often Xerox copies are not suitable for reproduction. For the same reason, any writing within figures should not be smaller than 2.5 mm. It might be useful to look at some of the volumes already published or, especially if some atypical text is planned, to write to the Physics Editorial Department of Springer-Verlag direct. This avoids mistakes and time-consuming correspondence during the production period.
As a special service, we offer free of charge LaTeX and TeX macro packages to format the text according to Springer-Verlag's quality requirements. We strongly recommend authors to make use of this offer, as the result will be a book of considerably improved technical quality.
For further information please contact Springer-Verlag, Physics Editorial Department II, Tiergartenstrasse 17, D-69121 Heidelberg, Germany.

Series homepage – http://www.springer.de/phys/books/lnpm

Piotr T. Chruściel Jacek Jezierski Jerzy Kijowski

Hamiltonian Field Theory in the Radiating Regime

Springer

Authors

Piotr T. Chruściel
Département de Mathématiques, Faculté des Sciences
Université de Tours
Parc de Grandmont, 37200 Tours, France

Jacek Jezierski
Department of Mathematical Methods in Physics
University of Warsaw
ul. Hoza 74, 00-682 Warsaw, Poland

Jerzy Kijowski
Center for Theoretical Physics, Polish Academy of Sciences
al. Lotników 32/46, 02-668 Warsaw, Poland

Library of Congress Cataloging-in-Publication Data applied for.

Die Deutsche Bibliothek - CIP-Einheitsaufnahme
Chruściel, Piotr T.: Hamiltonian field theory in the radiating regime / Piotr T. Chruściel ;
Jacek Jezierski ; Jerzy Kijowski. - Berlin ; Heidelberg ; New York ;
Barcelona ; Hong Kong ; London ; Milan ; Paris ; Tokyo : Springer, 2002
(Lecture notes in physics : N.s. M, Monographs ; 70)
(Physics and astronomy online library)
ISBN 3-540-42884-4

ISSN 0940-7677 (Lecture Notes in Physics. Monographs)
ISBN 3-540-42884-4 Springer-Verlag Berlin Heidelberg New York

Springer-Verlag Berlin Heidelberg New York
a member of BertelsmannSpringer Science+Business Media GmbH

http://www.springer.de

Typesetting: Camera-ready by the authors
Cover design: *design & production*, Heidelberg

Printed on acid-free paper
SPIN: 10857027 55/3141/du - 5 4 3 2 1 0

Contents

5.6 Energy in the radiation regime: convergence of integrals 89
5.7 Phase spaces: the space $\mathscr{P}$ 97
5.8 The phase space $\mathscr{P}_{[-1,0]}$ 102
5.9 The phase space $\widehat{\mathscr{P}}_{[-1,0]}$ 103
5.10 Preferred role of the Trautman–Bondi energy 104

6. Hamiltonians associated with the BMS group 105
6.1 The Poincaré group: convergence of integrals 105
6.2 Supertranslations (and space translations): convergence of integrals 111
6.3 The abstract Scri 116
6.4 Lorentz charges 119
6.5 A Hamiltonian definition of angular momentum of sections of $\mathscr{I}$ 121
6.6 An example: Schwarzschild space-time 126
6.7 An example: stationary space-times 129
6.8 Lorentz covariance of global charges 131
6.9 BMS invariance of energy-momentum 134
6.10 Polyhomogeneous Scri's 136

A. Odd forms (densities) 139

B. Solutions of the wave equation smoothly extendable to $\mathscr{I}^+$ 141

C. Gravitational field: some auxiliary results 143
C.1 The canonical gravitational variables in Bondi coordinates ... 143
C.1.1 Smooth Scri's 143
C.1.2 Polyhomogeneous asymptotics 145
C.2 Solutions of the vacuum Einstein equations containing hyperboloidal hypersurfaces 146
C.3 Bondi coordinates *vs* hyperboloidal initial data 152
C.4 The calculation of the $\mathbb{W}^{x^\mu}$'s 157
C.5 Transformation rules of the Bondi functions under supertranslations 159
C.6 Transformation rules of the Bondi functions under boosts 162
C.7 Bondi coordinates in the Kerr space-time 163
C.8 Conformal rescalings of ADM Cauchy data 166

References 167
[1]

---

[1] **Acknowledgements** JJ wishes to thank the Région Centre for financial support, and the Department of Mathematics of the Tours University for hospitality during part of work on this paper. We are grateful to M. MacCallum for help with a symbolic algebra calculation.

1. Introduction

In any physical theory one of the fundamental notions is that of energy of the objects at hand: in mechanics one considers the energy of, say, moving masses; in field theories one is interested in the energy of field configurations. A unified treatment of this question, which applies both to mechanics and to field theory, proceeds through a Hamiltonian formalism. We will shortly review below how such a procedure is carried out in the theory of scalar fields on Minkowski space-time; let us, at this stage, mention that an important issue, often ignored in the textbooks, is that of the boundary conditions satisfied by the set of fields under consideration. While this issue can be safely ignored — for many purposes — when considering the usual field theories, such as scalar fields or electromagnetism, on the $\{t = \text{const}\}$ hypersurfaces, where t is a Minkowski-time, it sometimes plays a critical role when other classes of hypersurfaces are considered. In the case of gravity the situation is worse: even for $\{t = \text{const}\}$ asymptotically Minkowskian slices the boundary terms are crucial. (This is one of the main differences between the Arnowitt-Deser-Misner (ADM) mass for gravity (*cf.* Sect. 5.4 below), which is given by a boundary integral, and the usual energy expression for field theories in Minkowski space-time, where the Hamiltonian is usually a volume integral.) Now, in field theory the energy plays its most important role in the radiation regime, where it can be radiated away by the field. This leads one to the need of considering hypersurfaces which extend to the radiation zone; this requirement is made precise by considering hypersurface which asymptote to null hypersurfaces in an appropriate way. (Technically, this will correspond to hypersurfaces with specific boundary behaviour in a conformal compactification of Minkowski space-time. In such compactifications the radiation zone becomes a neighbourhood of a conformal boundary $\mathscr{I}$.) The aim of this work is to analyse the issues which arise when attempting to obtain a Hamiltonian description of radiating fields, with emphasis on the geometric character of the objects involved. More precisely, we develop a geometric Hamiltonian formalism adequate for a canonical description of field theories in the radiation regime, extending previous work of Kijowski and Tulczyjew [95]. While our main objective here is the radiation zone, we note that several aspects of our construction are new even in more standard contexts. The formalism is first applied in detail to the toy model of a massless scalar field in Minkowski space-time at null infinity. This has some interest of its own; more importantly, it allows us also to adress the difficulties which arise in a simpler setting. Our real interest is the gravitational field, and we apply, next, our formalism to general relativity at null infinity. In particular we derive Hamiltonian formulae for energy, momentum, angular-momentum, as well as for the Hamiltonians for boosts and "supertranslations". Now, the — generally accepted — notion of energy in general relativity appropriate in the radiation regime is the one which has been introduced by Trautman [121], and further studied by Bondi [25]; we will refer to this mass as the Trautman–Bondi mass; the original motivation for this work was to show how this quantity arises in a Hamiltonian framework. One of the main results of the work here is a natural Hamiltonian definition of global Lorentz charges — that is, angular momentum and boost integrals — for cuts of $\mathscr{I}$, which is free from the "supertranslation ambiguities".

We shall now expand the quick overview, just given, of our work here. Let us start with a brief review of the Hamiltonian description of the dynamics of the massless scalar wave equation on Minkowski space-time,

$$\Box\phi = 0 \,. \tag{1.1}$$

Consider the collection of solutions of (1.1) with initial data which are, say, smooth and compactly[1] supported on the hypersurface $\mathscr{S}_0 = \{x^0 = 0\} \subset \mathbb{R}^{1,3}$, where $\mathbb{R}^{1,3}$ stands for the four dimensional Minkowski space-time. As is well known (and discussed in more detail below), this theory can be described as a dynamical system by considering the restrictions (ϕ_t, π_t) of $(\phi, \partial\phi/\partial t)$ to the hypersurfaces $\mathscr{S}_t = \{x^0 = t\}$. In this approach the family (ϕ_t, π_t), $t \in \mathbb{R}$, can be thought of as a smooth curve in the set $C_0^\infty(\mathbb{R}^3) \oplus C_0^\infty(\mathbb{R}^3)$ of smooth compactly supported functions on $\mathbb{R}^3$. The associated dynamical system is Hamiltonian, and in the standard formulation all the Hamiltonians generating the equations of motion are of the form

$$\mathscr{H} = \frac{1}{2}\int_{\mathbb{R}^3} (\pi^2 + |\nabla\phi|^2) d^3x + C \,. \tag{1.2}$$

This follows from the facts that: 1) $\mathscr{H}$ given by (1.2) is differentiable and satisfies the appropriate generating equations (*cf.* Sect. 2.1 below); 2) the difference of any two Hamiltonians has vanishing differential, and therefore must be a constant because the space of field configurations is path connected.

The constant in (1.2) can be gotten rid of by requiring that the energy vanishes for the trivial configuration $\phi = \pi = 0$. This requirement leads then to a uniquely defined quantity, usually identified with the total energy contained in a field configuration.

[1] The condition of smooth compactly supported initial data is made only for simplicity, and can be considerably relaxed. In particular, later on in this work, we shall consider field configurations which do not satisfy this hypothesis.

When one attempts to replace the hypersurfaces $\mathscr{S}_t$ above with hypersurfaces which extend to null infinity, various difficulties arise (they will be presented and adressed in the work below). In this context some Hamiltonian formulations of the dynamics have been presented [9, 12, 13, 74, 83], and in all those formulations the Hamiltonian turns out to be the energy calculated "at spatial infinity". More precisely, in the above mentioned descriptions of a scalar field in Minkowski space-time the Hamiltonian is, essentially[2], the energy calculated on hypersurfaces $\{x^0 = \text{const}\}$ as in Equation (1.2), where x^0 refers to the Minkowskian time coordinate. (Similarly, in the analysis of the gravitational field in [9, 12, 13, 74, 83] the Hamiltonian is, essentially, the ADM mass.) Now we are interested in a definition of the energy in the radiating regime, where one expects that the correct energy should *not* be given by the integrals of the energy-momentum tensor on the level sets of the Minkowskian time, but on hypersurfaces extending into the radiation zone; this would be the scalar field equivalent of the Trautman–Bondi mass in general relativity [25, 121], and we will use those names when refering to that mass. It has been argued [126] that no such formulation is possible, because in a Hamiltonian system the energy is conserved, while the Trautman–Bondi mass is not. We shall show that the argument of [126] does not apply when things are suitably formulated, and that there exists a Hamiltonian description of the dynamics of the scalar field in the radiating regime in which the Hamiltonian is the Trautman–Bondi mass; see Sects. 2.4 and 2.5 for a simple exposition of these ideas.

Recall, next, that in a manner rather analogous to the scalar field on Minkowski space-time, the Einstein equations induce a dynamical system on the phase space of those gravitational initial data which are asymptotically flat in spacelike directions; this is discussed in more detail in Sect. 5.4 below. In this case all the Hamiltonians corresponding to space-time motions which reduce to unit time translations to the future in the asymptotically flat regions, are of the form

$$\mathscr{H} = M_{ADM} + C \,,$$

$$M_{ADM} = \frac{1}{16\pi} \int_{S_\infty} (g_{ij,j} - g_{jj,i}) dS_i \tag{1.3}$$

(the integral over the union "S_∞" of all the "spheres at infinity" is understood as a limit as R tends to infinity of integrals over the union of spheres of radius R in all the asymptotically flat regions), with the constant C usually set to

[2] The qualification "essentially" here is due to the fact that the equality of the Hamiltonian with the energy of Minkowskian-time slices $\{x^0 = \text{const}\}$ is correct for smooth compactly supported initial data, and is expected to be true for the more general data considered in those works. However, no such rigourous results are known even in the case of the scalar field on Minkowski space-time. A similar comment applies to the gravitational field.

zero[3]. Here g_{ij} denotes the metric on the "$t = 0$ hypersurface", with indices i, j in (1.3) running from 1 to 3, and being summed over.

In complete analogy to the scalar field case the situation in the radiation regime is much less satisfactory. Here one expects the Trautman–Bondi energy [25, 109, 120, 121] to be the physically relevant measure of the total energy contained in a hypersurface $\mathscr{S}$ that intersects $\mathscr{I}^+$ in an appropriate way. To our knowledge no satisfactory theoretical justification of such a statement has been given so far, though some partial results can be found in [24, 43, 126]. The strongest hint in this direction seems to be given by the uniqueness theorem of [43], which asserts that the Trautman–Bondi energy is the unique functional, up to a multiplicative constant, in an appropriate class of functionals, which is monotonic in time under time translations of $\mathscr{S}$. While monotonicity is certainly a reasonable condition, it is not clear to us that the requirement of monotonicity is a sufficient criterion for excluding all other possibilities. We emphasize that this problem has nothing to do with the gravitational field, as it occurs already for a massless scalar field in Minkowski space-time.

The purpose of this monograph is to show that there exists a Hamiltonian description of dynamics of a massless scalar field, as well as of the dynamics of the gravitational field, in the radiation regime. We construct such a framework, and exhibit two different ways in which the Trautman–Bondi energy arises. The first such occurrence is by taking an appropriate limit of the Hamiltonians on the phase space $\widehat{\mathscr{P}}_{[-1,0]}$ of Sects. 4.5 or 5.9 below. This gives a unique result, up to one normalization constant, on each connected component of the phase space. Next, we show that the Trautman–Bondi energy is one of the Hamiltonians on yet another phase space (the phase space $\mathscr{P}_{[-1,0]}$ of Sects. 4.4 or 5.8). For reasons which we discuss in detail below the freedom of adding a constant to any Hamiltonian leads to essential ambiguities, related to the nature of $\mathscr{P}_{[-1,0]}$, which we describe in an exhaustive way. While those ambiguities are somewhat reminiscent of the ones that arise in the "Noether charge" approach (*cf.*, *e.g.*, [30, 43]), the arbitrariness left turns out to be considerably smaller. We are unaware of any natural prescription which would remove that arbitrariness. We give arguments, parallel to those in [43], which indicate that the requirement of monotonicity with respect to time translations to the future singles out a unique Hamiltonian — the Trautman–Bondi energy. The analysis here is similar to, but not identical with the one in [43], because we work in the class of smoothly conformally compactifiable space-times which have complete spacelike hypersurfaces. We note that the class of functionals, which are Hamiltonians, is considerably

[3] While the choice $C = 0$ is rather reasonable, it is not clear whether this is the best one in all situations: the topology of the initial surface can be varied at will, so that the space of initial data is certainly not connected in any reasonable topology, and one is free to choose different values of C on different connected components of the phase space. This freedom could have some physical significance, *e.g.* when a path integral is performed.

smaller than the class of the functionals considered in [43], but the problem here is more constrained in view of the global conditions imposed.

It should be pointed out that some of our constructions are somewhat related to those of [9]. Those authors consider fields which, when extended to $\mathscr{I}^+$, are defined on the semi-global sets $\mathscr{I}^+_{(-\infty,\tau)}$, *cf.* Equation (4.20) below. As already pointed out, in the approach of [9] the dynamics is (up to severe mathematical difficulties) Hamiltonian, with the Hamiltonian equal to (again ignoring some mathematical problems) the ADM mass, and *not* the Trautman–Bondi mass. Moreover, our approach allows us to avoid altogether those difficulties [9], which are related to global existence questions for the general relativistic Cauchy problem, as well as to convergence of various integrals on $\mathscr{I}^+_{(-\infty,\tau)}$.

This work is organized as follows: In Chap. 2 we start by recalling some elementary facts concerning Hamiltonian dynamical systems, and we give some toy examples illustrating some of the ideas developed in the remainder of this work. In Chap. 3 we describe our geometric Hamiltonian framework, adequate both to the usual asymptotically-flat-at-spatial-infinity regime and to the radiation regime, which generalizes the framework of [95]. We note that our framework clarifies some questions which arise already in standard contexts, in particular the question of interpretation of general relativistic initial data sets with vanishing lapse function. As far as the radiation zone is concerned, it turns out that the case of the massless scalar field on Minkowski space-time already exhibits several essential features of the problems that arise there, while avoiding various technicalities which occur when one wishes to describe the Einstein gravity. Therefore we continue, in Chap. 4, with a detailed description of the application of our formalism to the case of the massless scalar field. The formalism of Chap. 3 is applied to the case of Einstein gravity in Chaps. 5 and 6. The inspection of the table of contents should give the reader a faithful impression of the contents of the various sections.

2. Preliminaries

2.1 Hamiltonian dynamics

There exist different approaches to the definition of a Hamiltonian system (*cf.*, *e.g.*, [2, 32, 95, 98]). An exhaustive treatment in the infinite dimensional case would involve delicate considerations concerning the manifold structure of the spaces at hand; in particular, one would have to introduce the notion of tangent vectors, differential forms, as well as an appropriate notion of non-degeneracy and closedness of the symplectic form. We do not wish to enter into such questions, and the purpose of this chapter is to present a simple minded approach which avoids all those issues. We shall show in the following chapters that the dynamics of the massless scalar field in the radiation regime, as well as that of the gravitational field in the radiation regime, satisfies the requirements of the definitions given in this chapter.

We begin with the definition of an *autonomous* Hamiltonian system, where both the dynamics and the associated Hamiltonian function H are time-independent. Consider a vector space $\mathscr{P}$ in which the notion of a differentiable curve can be defined. (In this work, unless indicated otherwise, when $\mathscr{P}$ is a space of differentiable functions on a set $\mathscr{O}$, then we will say that a curve $f_\lambda \in \mathscr{P}$, $\lambda \in \mathbb{R}$, is differentiable if the family of functions f_λ is jointly differentiable in $\lambda \in \mathbb{R}$ and $x \in \mathscr{O}$.) Recall that a family of maps $T_\lambda : \mathscr{P} \to \mathscr{P}$, $\lambda \in \mathbb{R}$, is called a differentiable dynamical system on $\mathscr{P}$ if

1. $T_0 = \mathrm{id}$, the identity map on $\mathscr{P}$, if
2. $T_a \circ T_b = T_{a+b}$, and if
3. for every $p \in \mathscr{P}$ the orbit of T_λ defined as the map $\lambda \to T_\lambda(p) \in \mathscr{P}$ is a differentiable curve on $\mathscr{P}$.

We set

$$\mathfrak{X}(p) = \left.\frac{dT_\lambda(p)}{d\lambda}\right|_{\lambda=0} . \tag{2.1}$$

We will call $\mathfrak{X}$ *a vector field generating the dynamics*; by this we *only* mean that equation (2.1) holds. In particular, no hypotheses are made about the possibility of recovering T_λ from $\mathfrak{X}$ — here the fundamental object is T_λ.

Let Ω be a bilinear antisymmetric map on $\mathscr{P}$ with values in $\mathbb{R}$; the Ω's we will consider will satisfy some non-degeneracy conditions but we do not need to make those precise here. We shall say that the dynamical system is

Hamiltonian if there exists a function H on $\mathscr{P}$ such that for all differentiable curves p_σ on $\mathscr{P}$ we have

$$\frac{dH(p_\sigma)}{d\sigma}\Big|_{\sigma=0} = -\Omega\left(\mathfrak{X}, \frac{dp_\sigma}{d\sigma}\Big|_{\sigma=0}\right) . \tag{2.2}$$

(We note that this definition implicitly requires $H(p_\sigma)$ to be differentiable at $\sigma = 0$ whenever p_σ is.) The function H will be called *a Hamiltonian* for the dynamical system $(\mathscr{P}, T_\lambda)$.

As an illustration, let $\mathscr{P} = C_0^\infty(\mathbb{R}^3) \oplus C_0^\infty(\mathbb{R}^3)$ be the space of pairs of smooth compactly supported functions on $\mathbb{R}^3$. Let f be a solution of the massless scalar field equation on Minkowski space-time,

$$\Box f = 0 ,$$

satisfying

$$f(t=0) = \varphi , \qquad \frac{\partial f}{\partial t}(t=0) = \pi , \qquad (\varphi, \pi) \in \mathscr{P} ,$$

and set

$$\mathscr{P} \ni (\varphi, \pi) \to T_\lambda(\varphi, \pi) = \left(f(t=\lambda), \frac{\partial f}{\partial t}(t=\lambda)\right) \in \mathscr{P} .$$

Here t is a Minkowskian time coordinate in Minkowski space-time. If we equip $\mathscr{P}$ with the standard ("symplectic") form,

$$\Omega((\varphi_1, \pi_1), (\varphi_2, \pi_2)) = \int_{\mathbb{R}^3} (\varphi_1 \pi_2 - \varphi_2 \pi_1) d^3x ,$$

then the dynamical system (P, T_λ) is Hamiltonian in the above sense, with

$$H(\varphi, \pi) = \frac{1}{2} \int_{\mathbb{R}^3} (|D\varphi|^2 + \pi^2)\, d^3x . \tag{2.3}$$

Here $|D\varphi|$ denotes the length of the space-gradient of φ.

It turns out that even for the massless scalar field we need to generalize the set-up above, allowing *non-autonomous* (time-dependent) Hamiltonian systems. More precisely,

1. it will be necessary to consider a dynamical system generated (in the sense of equation (2.5) below) by a time-dependent "vector field" $\mathfrak{X}_t$,
2. with the corresponding time-dependent flow $T_{t,s}$ only locally defined.

More precisely, we will consider a family $T_{t,s}$ of maps

$$\mathbb{R} \times \mathbb{R} \times \mathscr{P} \supset \mathscr{U} \ni (t, s, p) \to T_{t,s}(p) \in \mathscr{P} , \tag{2.4}$$

defined on an open connected subset $\mathscr{U}$ of $\mathbb{R} \times \mathbb{R} \times \mathscr{P}$. Those maps describe where a trajectory of the dynamical systems passing through a point p at time s will arrive at time t. We shall further require:

1. For all $p \in \mathscr{P}$ the maps $T_{t,s}$ are defined on an open set of t's and s's, containing zero: $\mathscr{U} \supset \{0\} \times \{0\} \times \mathscr{P}$.
2. The composition formula
$$T_{t_3,t_2} \circ T_{t_2,t_1} = T_{t_3,t_1}$$
holds whenever all the objects in the above equation are simultaneously defined.
3. For all $(t,p) \in \mathbb{R} \times \mathscr{P}$ the curves $s \to T_{t+s,t}(p)$ are differentiable at $s = 0$.

We set
$$\mathfrak{X}_t(p) = \left.\frac{dT_{t+s,t}}{ds}\right|_{s=0} . \tag{2.5}$$

We mention that in one of the cases considered below for the massless scalar field (the phase space $\widehat{\mathscr{P}}_{[-1,0]}$ of Sect. 4.5) the set $\mathscr{U}$ will be of the form
$$\mathscr{U} = \{t \in (-1,\infty), s \in (-1,\infty), p \in \mathscr{P}\} . \tag{2.6}$$

In the scalar field case the restrictions on t and s that follow from (2.6) arise because we will mainly be interested in those solutions of the massless scalar field equation which are defined to the future of a given hyperboloid in Minkowski space-time. In the gravitational field case there is a further fundamental reason for allowing a $\mathscr{U}$ not necessarily equal to $\mathbb{R} \times \mathbb{R} \times \mathscr{P}$, related to the blow up in finite time of solutions of the Einstein equations.

The equivalent of (2.2) reads
$$\left.\frac{dH(t,p_\sigma)}{d\sigma}\right|_{\sigma=0} = -\Omega\left(\mathfrak{X}_t, \left.\frac{dp_\sigma}{d\sigma}\right|_{\sigma=0}\right) , \tag{2.7}$$

for all curves p_σ differentiable at $\sigma = 0$, and for those t for which $(t, 0, p = p_0) \in \mathscr{U}$ (so that $\mathfrak{X}_t$ is defined). Equation (2.7) is the desired generalization of the notion of a Hamiltonian dynamical system to the time-dependent case, with the Hamiltonian H being defined on an appropriate subset of $\mathbb{R} \times \mathscr{P}$.

Recall that there is another standard way of dealing with time-dependent Hamiltonians [98, Chap. V, p. 328], which consists in enlarging the phase space by adding to it t and its conjugate variable p^0. While this can be done in our case, we have found the approach above to be simpler.

The global time-independent formulation is a special case of the local time-dependent one if one sets $T_{t,s} \equiv T_{t-s}$, $\mathscr{U} \equiv \mathbb{R} \times \mathbb{R} \times \mathscr{P}$.

Let us close this section with the simple observation, that functionals satisfying equation (2.7) are unique up to a constant *when the phase space* $\mathscr{P}$ *is connected, and locally path connected* via differentiable paths. Indeed, if H_1 and H_2 satisfy equation (2.7) then $H_1 - H_2$ has vanishing derivative along any one-differentiable one-parameter family of fields, so that $H_1 - H_2$ is locally constant by local path connectedness, hence constant by connectedness of $\mathscr{P}$.

2.2 The role of boundary conditions in Hamiltonian field theory

In this section we want to give a short and informal overview of the ideas, which we later use to describe radiation phenomena in the Hamiltonian field theory. Let us analyse more in detail the definition of the Hamiltonian flow, given by formula (2.7), in the case of the scalar wave equation. We introduce the following notation: whenever we have a differentiable family of functions $f(x;\sigma)$, where the variables x describe the position of a point in the physical space, or in space-time, and σ is an abstract parameter, then the derivative of this family with respect to the parameter, calculated at $\sigma = 0$ will be denoted by δ:

$$\delta f(x) := \left.\frac{\partial f(x;\sigma)}{\partial \sigma}\right|_{\sigma=0} . \tag{2.8}$$

This notation is standard in the calculus of variation. In functional spaces considered in this monograph (as, *e.g.*, in the space $\mathscr{P} = \{(\varphi, \pi)\}$ of Cauchy data for the wave equation, considered in the previous section), *differentiable curves* are simply differentiable one-parameter families of functions. Whenever we meet a "variation of a function", we understand that a one-parameter family $f(x;\sigma)$ of functions over physical space (or space-time) has been chosen and the derivative (2.8) has been calculated. In this notation, the left hand sides of (2.2) or (2.7) become simply δH.

Another derivative, which we shall often use in our work, is the Lie derivative $\mathscr{L}_X$ or (in the case of field theories which are more general than the scalar field theory) the covariant derivative $\mathscr{D}_X$, along a space-time vector field X defining the evolution which we want to describe. It is sometimes convenient to use coordinates adapted to this vector field so that we have $X = \partial_0$, and the corresponding Lie derivative reduces to the time derivative. Whenever it does not lead to any misunderstanding, we shall denote it by a "dot". As will be seen in the next chapter, the use of adapted coordinates may be avoided and the entire Hamiltonian field theory formulated in geometric terms, both for flows of vector fields and for motions of hypersurfaces.

In adapted coordinates used in the previous section, the components of the vector (2.5), defined on the space of Cauchy data $\mathscr{P} = \{(\varphi, \pi)\}$ for the scalar field theory are simply denoted by $(\mathscr{L}_X\varphi, \mathscr{L}_X\pi) = (\dot{\varphi}, \dot{\pi})$, where a dot denotes now the usual derivative with respect to the Minkowskian time coordinate, in a Minkowskian coordinate system. The field equation

$$\Box f = \Delta f - \ddot{f} = 0 ,$$

expressed in terms of Cauchy data (φ, π), gives the following system of equations:

$$\dot{\varphi} = \pi , \tag{2.9}$$

$$\dot{\pi} = \Delta\varphi . \tag{2.10}$$

Using the definition of the form Ω in space $\mathscr{P}$, given in the previous section, we may rewrite the equation (2.7) for the scalar field in the following way:

$$\begin{aligned} -\delta H &= \int_{\mathbb{R}^3} (\dot{\pi}\delta\varphi - \dot{\varphi}\delta\pi) d^3x \\ &= \int_{\mathbb{R}^3} ((\mathscr{L}_X\pi)\delta\varphi - (\mathscr{L}_X\varphi)\delta\pi) d^3x \ . \end{aligned} \tag{2.11}$$

Let us analyse in more detail the mechanism which leads to this equation. For this purpose we calculate explicitly the variation of the Hamiltonian (2.3). Because derivatives with respect to the parameter σ commute with derivatives with respect to space variables, we have:

$$\begin{aligned} -\delta H(\varphi,\pi) &= -\delta \left\{ \frac{1}{2} \int_{\mathbb{R}^3} (|D\varphi|^2 + \pi^2)\, d^3x \right\} \\ &= - \int_{\mathbb{R}^3} (D\varphi \cdot D\delta\varphi + \pi\delta\pi)\, d^3x \\ &= \int_{\mathbb{R}^3} (\Delta\varphi\delta\varphi - \pi\delta\pi)\, d^3x - \int_{\mathbb{R}^3} D(D\varphi\delta\varphi)\, d^3x \ . \end{aligned} \tag{2.12}$$

Hence, equation (2.12) is equivalent to (2.11) if and only if the last integral vanishes. Due to the Stokes theorem, it may be converted into a "surface integral at infinity" of the vector field $-(D\varphi)\delta\varphi$. It vanishes if sufficiently fast fall-off conditions are imposed on the field f; for definiteness, as in the previous section we assume that f is compactly supported on each hypersurface of constant Minkowskian time, but much weaker asymptotic conditions are of course sufficient.

Now, consider the evolution of the same scalar field f but in a finite volume $V \subset \mathbb{R}^3$, with non-empty boundary. Let H_V be the total amount of the usual field energy contained in V:

$$H_V(\varphi,\pi) = \frac{1}{2} \int_V (|D\varphi|^2 + \pi^2)\, d^3x \ . \tag{2.13}$$

Similar calculations as above lead to the following result:

$$-\delta H_V(\varphi,\pi) = \int_V (\Delta\varphi\delta\varphi - \pi\delta\pi)\, d^3x + \int_{\partial V} (\pi^a\delta\varphi)\, d\sigma_a \ , \tag{2.14}$$

where we have introduced the notation

$$\pi^a := -D^a\varphi \ , \tag{2.15}$$

and used the Stokes theorem to convert the last integral into a surface integral. To recover the definition of a Hamiltonian (2.7) — or, equivalently, equation (2.11) — we must annihilate the surface integral by imposing some boundary conditions on the elements of our phase space $\mathscr{P} = \{(\varphi,\pi)\}$. It should be stressed that the need for imposing boundary conditions does not arise only because we wish to have a formula such as (2.7); without boundary conditions the time evolution does not define a dynamical system, since then the initial value problem is not well posed: the scalar field f in the future may be changed due to incoming or outgoing radiation, even if the Cauchy data at $t = 0$ remain the same. Boundary conditions are *necessary* to obtain a deterministic system. Physically, they can be thought of as our control of the radiation passing through the boundary ∂V of the region V, in which we perform experiments.

In the example just given the boundary of the space-time region $\mathbb{R} \times V$ is a timelike hypersurface. In such situations a simple way to annihilate the surface integral in (2.14) is to choose some function $\psi : \mathbb{R} \times \partial V \to \mathbb{R}$ and to impose on the scalar field f the Dirichlet boundary condition

$$f\big|_{\mathbb{R}\times\partial V} = \psi \,, \tag{2.16}$$

for all functions used in the sequel[1]. This implies

$$\varphi\big|_{\partial V} \equiv f\big|_{\{0\}\times\partial V} = \psi\big|_{\{0\}\times\partial V} \implies \delta\varphi\big|_{\partial V} = 0 \,, \tag{2.17}$$

within the class of functions satisfying condition (2.16). If ψ is time-independent, we obtain in this way a well defined, autonomous Hamiltonian system. For a time-dependent boundary condition one could think, at a first glance, that the very notion of a phase-space $\mathscr{P}_t = \{(\varphi, \pi) : \varphi\big|_{\partial V} = \psi(t, \cdot)\}$ does depend upon time and, therefore, no Hamiltonian description is possible. The remedy to this difficulty is, however, straightforward: choose any time-dependent function $\phi = \phi(t, x)$, which satisfies the boundary condition (2.16), and parameterize the data (φ, π) by the following functions:

$$\widetilde{\varphi} := \varphi - \phi \,, \qquad \widetilde{\pi} := \pi - \dot{\phi} \,. \tag{2.18}$$

The new variables fulfill a homogeneous, time independent, boundary condition

$$\widetilde{\varphi}\big|_{\partial V} = 0 \,, \tag{2.19}$$

and equations (2.18) provide an identification of all the phase spaces $\mathscr{P}_t$ with the time-independent phase space $\widetilde{\mathscr{P}} = \{(\widetilde{\varphi}, \widetilde{\pi}) : \widetilde{\varphi}\big|_{\partial V} = 0\}$. The Hamiltonian description of the field theory is, therefore, applicable (*i.e.*, formula (2.11) remains valid) also in case of time-dependent boundary data. The only price we pay for this is an explicit time-dependence of the Hamiltonian, arising from the (given *a priori*) "reference function" ϕ and its derivatives, when

[1] In the case of null boundaries, or for the description of radiation, this method does not apply; a simple example illustrating this will be given in Sect. 2.4 below. The point of the examples in this section is not to show in a simple case how we handle the radiation problem, but to give an indication of the kind of problems that arise when domains with boundary are considered.

we substitute $\varphi := \widetilde{\varphi} + \phi$ and $\pi := \widetilde{\pi} + \dot{\phi}$ in formula (2.13). Physically, the non-autonomous properties of the field dynamics within V is due to time-dependent external forces, applied on the boundary ∂V, in order to control the boundary data of the field in a prescribed way.

Imposing Dirichlet boundary conditions is *by no means* a unique way to annihilate the surface integral in (2.14), obtaining thus a Hamiltonian dynamical system. A frequently used alternative consists in imposing Neumann conditions, *i.e.*, prescribing the value of $\pi^a \, d\sigma_a$ on the boundary. For this purpose we write

$$\pi^a \delta\varphi = \delta(\pi^a \delta\varphi) - \varphi\delta\pi^a ,$$

and transfer the first term to the left-hand side of (2.14). (The manipulations involved are somewhat reminiscent of those which arise when Legendre transformations are carried on.) This leads us to the formula

$$-\delta \widetilde{H}_V(\varphi, \pi) = \int_V (\Delta\varphi\delta\varphi - \pi\delta\pi) \, d^3x - \int_{\partial V} (\varphi\delta\pi^a) \, d\sigma_a , \tag{2.20}$$

where the new Hamiltonian, describing the mixed Cauchy–Neumann evolution equals:

$$\widetilde{H}_V(\varphi, \pi) := H_V(\varphi, \pi) + \int_{\partial V} (\pi^a\varphi) \, d\sigma_a = H_V(\varphi, \pi) + \int_V D_a(\pi^a\varphi) \, d^3x . \tag{2.21}$$

equation (2.20) leads to the formula (2.11) for the Hamiltonian evolution of the Cauchy data, because the Neumann boundary condition on $\pi^a d\sigma_a$ implies:

$$\delta(\pi^a \, d\sigma_a)\big|_{\partial V} \equiv 0 ,$$

within the class of functions allowed by the condition and, therefore, the surface integral vanishes.

These issues will be discussed in following chapters under more general circumstances, and the mathematical structures associated with the above formulae will be described. One of the points of the examples given was to stress that in some situations there might be many ways to translate the field evolution into the language of Hamiltonian dynamics. Different physical situations lead to different boundary conditions. For the gravitational field, governed by Einstein equations, possible choices of boundary conditions on bounded domains with boundary have been discussed in [93], [92] and [94] from a symplectic point of view. It is expected that those considerations might shed some light on the associated analytic problem; see [65] for some rigorous analytic results that do not involve symplectic considerations.

2.3 Tangential translations as a Hamiltonian system

In the usual treatment of relativistic field theories on Minkowski space-time the field energy provides the Hamiltonian for the time evolution; here the dynamics is associated with the space-time vector field $X = \partial_t$, which is transversal to the Cauchy surfaces. Under some circumstances one might be interested in evolution of the fields under motions associated with vector fields tangent to the Cauchy surface; we will encounter such situations throughout this monograph. Again in textbook treatments, the "generator of tangential space translations" is the momentum of the field configuration. Let us analyse more carefully such a "dynamics" for motions of the massless scalar field under the group T_λ of space translations, generated by the field $X = \partial_a$:

$$\mathscr{P} \ni (\varphi, \pi) \to T_\lambda(\varphi, \pi) := (T_\lambda(\varphi), T_\lambda(\pi)) \in \mathscr{P} , \tag{2.22}$$

where

$$T_\lambda(\varphi)(x) := \varphi(x + \lambda e_a) , \tag{2.23}$$
$$T_\lambda(\pi)(x) := \pi(x + \lambda e_a) , \tag{2.24}$$

and e_a denotes the versor of a-th axis in $\mathbb{R}^3$. The vector field (2.5), corresponding to this evolution, equals:

$$(\mathscr{L}_X\varphi, \mathscr{L}_X\pi) = (\partial_a\varphi, \partial_a\pi) .$$

Let us check that the a-th component of the field momentum,

$$P_a(\varphi, \pi) = \int_{\mathbb{R}^3} (\pi\partial_a\varphi)\, d^3x , \tag{2.25}$$

provides indeed a Hamiltonian for the dynamics, by calculating its variation:

$$-\delta P_a(\varphi, \pi) = -\int_{\mathbb{R}^3} (\pi\partial_a\delta\varphi + (\partial_a\varphi)\delta\pi)\, d^3x \tag{2.26}$$
$$= \int_{\mathbb{R}^3} ((\partial_a\pi)\delta\varphi - (\partial_a\varphi)\delta\pi)\, d^3x - \int_{\mathbb{R}^3} \partial_a(\pi\delta\varphi)\, d^3x \tag{2.27}$$
$$= \int_{\mathbb{R}^3} ((\mathscr{L}_X\pi)\delta\varphi - (\mathscr{L}_X\varphi)\delta\pi) d^3x . \tag{2.28}$$

The last equation is satisfied, because the integral of a total divergence vanishes when sufficiently fast asymptotic fall-of conditions are imposed on φ; recall that we are assuming that φ is compactly supported. In the case of a bounded domain V, the resulting boundary integral

$$I_{\partial V} = \int_{\partial V} (\pi\delta\varphi)\, d\sigma_a ,$$

provides in general an obstruction to the Hamiltonian character of the associated dynamics.

Replacing ∂_a by an an arbitrary complete vector field $X = X^a\partial_a$, tangent to the Cauchy space $\mathbb{R}^3$, and the group of rigid space translations by the one-parameter group of diffeomorphisms $\mathscr{G}^X$ generated by X, we may generalize the above example. A "time evolution" associated with $\mathscr{G}^X$ can be defined by Lie transporting both φ and π along the flow of the field X; we stress that the "time" involved has nothing to do with the physical time, and is simply a parameter along the integral curve of X. With this definition the vector tangent to the evolution curve is given by the Lie derivative of both objects with respect to X:

$$(\mathscr{L}_X\varphi, \mathscr{L}_X\pi) = (X^a\partial_a\varphi, \partial_a(X^a\pi)) \, . \tag{2.29}$$

(The last expression for the Lie derivative of π is due to the fact that the momentum is not a scalar function but a scalar density. This is discussed in more detail in the next chapter.)

Let P_V be given by the formula

$$P_V^X(\varphi, \pi) := \int_V (\pi X^a \partial_a \varphi) \, d^3x \, ; \tag{2.30}$$

calculating its variation we obtain:

$$-\delta P_V^X(\varphi, \pi) = \int_V ((\mathscr{L}_X\pi)\delta\varphi - (\mathscr{L}_X\varphi)\delta\pi)d^3x + \int_{\partial V} (\pi^a\delta\varphi) \, d\sigma_a \, , \tag{2.31}$$

where we have set

$$\pi^a := -X^a\pi \, . \tag{2.32}$$

The dynamics associated with the dragging of the scalar field along a vector field X is, again, Hamiltonian if the surface integral vanishes. This occurs without the need of imposing any boundary conditions on the initial data when X is tangent to the boundary ∂V: in such a case the surface integral vanishes identically and P_V^X becomes a Hamiltonian. This holds *e.g.*, for the one-parameter group of rotations, whenever they leave V invariant. In that case, P_V^X is usually identified with the total amount of angular momentum carried by the field within V.

2.4 The Hamiltonian description of a mixed Cauchy – characteristic initial value problem

The main purpose of this monograph is to give a description of radiation phenomena in terms of Hamiltonian dynamics. Those phenomena are best captured by adding to space-time a conformal boundary, called *Scri*, and denoted by the symbol $\mathscr{I} = \mathscr{I}^+ \cup \mathscr{I}^-$; outgoing radiation can then be studied in a neighbourhood of $\mathscr{I}^+$, while ingoing radiation is related to the behaviour of the fields in a neighbourhood of $\mathscr{I}^-$. $\mathscr{I}^+$ is a null-like, three-dimensional manifold, with structure similar to that of a light cone. To illustrate methods which will be used to describe field dynamics on $\mathscr{I}^+$, let us consider a toy example, in which the "*asymptotic* light cone $\mathscr{I}^+$" is replaced by a standard, finite light cone. Now, a Hamiltonian description of the field dynamics within the future-oriented light cone

$$\mathscr{C}^+ = \{(t,x) : \|x\| < t\}$$

must take into account the incoming radiation, which enters $\mathscr{C}^+$ through its boundary. In this monograph we have concentrated on a description of the outgoing radiation. Hence, we use the past oriented cone $\mathscr{C}^- = \{(t,x) : \|x\| < -t\}$, to make our toy model better adapted to this purpose. Both cases are, however, symmetric: replacing t with $-t$ one obtains a toy model for outgoing radiation from the incoming radiation one, and vice versa.

To eliminate technicalities and make the model as simple as possible, let us restrict ourselves to the two-dimensional Minkowski space. This means that space is one dimensional: $x \in \mathbb{R}^1$. We consider again a massless scalar field, solving the wave equation. We want to describe its Cauchy data on the surfaces $\{t = \text{const}\}$ in the interior of the cone $\mathscr{C}^-$. To be able to identify these surfaces for different times, let us introduce new coordinates $(\xi^\mu) = (\tau, \xi)$ (where $\mu = 0, 1$), related to the Minkowskian coordinates $(x^\mu) = (t, x)$ in the following way:

$$t = -e^{-\tau} \, , \tag{2.33}$$

$$x = \xi e^{-\tau} \, , \tag{2.34}$$

where $\tau \in \mathbb{R}^1$ and $|\xi| \leq 1$. Within this range, the new coordinates parameterize the entire cone $\mathscr{C}^-$. To derive the Hamiltonian description of the wave equation in these coordinates, we use the textbook procedure, based on the standard, relativistic-invariant Lagrangian

$$\mathbf{L} = L \, d^2x \, , \tag{2.35}$$

where

$$L = -\frac{1}{2} g^{\mu\nu}(\partial_\mu f)(\partial_\nu f) = \frac{1}{2}\left\{(\partial_t f)^2 - (\partial_x f)^2\right\} \, . \tag{2.36}$$

We rewrite this Lagrangian in the coordinates (τ, ξ), using the following formulae which may be easily derived from (2.33) and (2.34):

$$\partial_t = e^\tau \left(\partial_\tau + \xi \partial_\xi\right) \, ,$$

$$\partial_x = e^\tau \partial_\xi \, .$$

Moreover, we have

$$d^2x = dt dx = e^{-2\tau} d\tau d\xi = e^{-2\tau} d^2\xi \, .$$

Expressing the Lagrangian (2.35) in terms of new coordinates we thus obtain:

$$\mathbf{L} = \mathscr{L} \, d^2\xi \, , \tag{2.37}$$

where

$$\mathscr{L} = \frac{1}{2}\left\{ (\partial_\tau f + \xi \partial_\xi f)^2 - (\partial_\xi f)^2 \right\} . \tag{2.38}$$

The standard procedure, valid for an arbitrary Lagrangian density $\mathscr{L} = \mathscr{L}(f, f_\mu, \xi^\mu)$, where $f_\mu := \partial_\mu f$, proceeds as follows: We introduce generalized momenta:

$$\pi^\mu := \frac{\partial \mathscr{L}}{\partial f_\mu} , \tag{2.39}$$

and calculate the variation of the Lagrangian:

$$\delta \mathscr{L} = \frac{\partial \mathscr{L}}{\partial f} \delta f + \pi^\mu \delta f_\mu = \partial_\mu \left(\pi^\mu \delta f \right) + \left(\frac{\partial \mathscr{L}}{\partial f} - \partial_\mu \pi^\mu \right) \delta f . \tag{2.40}$$

The field equation $\Box f = 0$ is equivalent to the vanishing of the Euler–Lagrange term in (2.40):

$$\frac{\partial \mathscr{L}}{\partial f} - \partial_\mu \pi^\mu = 0 , \tag{2.41}$$

and, therefore, is equivalent to the following equation which must be fulfilled by the variation of $\mathscr{L}$:

$$\delta \mathscr{L} = \partial_\mu \left(\pi^\mu \delta f \right) = \left(\pi \delta \varphi \right)^{\cdot} + \partial_\xi \left(\pi^1 \delta \varphi \right) , \tag{2.42}$$

where we have denoted by φ the restriction of the field f to the Cauchy surface $\Sigma = \{\tau = \text{const.}\}$ and by a dot — the derivative with respect to the new time variable τ. Moreover, we have introduced the momentum $\pi := \pi^0$, which provides the remaining piece of Cauchy data on the surface[2]. Integrating the field equation (2.42) over a volume V in the Cauchy surface $\Sigma = \{\tau = \text{const.}\}$, we obtain the following identity, valid for fields satisfying the wave equation:

$$\begin{aligned} \delta \int_V \mathscr{L} \, d\xi &= \int_V \left(\pi \delta \varphi \right)^{\cdot} \, d\xi + \int_{\partial V} \left(\pi^1 \delta \varphi \right) \, d\sigma_1 \\ &= \int_V \left(\dot{\pi} \delta \varphi - \dot{\varphi} \delta \pi + \delta(\pi \dot{\varphi}) \right) \, d\xi + \left[\pi^1 \delta \varphi \right]_{\partial V} , \end{aligned} \tag{2.43}$$

where the integral over the 0-dimensional boundary ∂V is equal to the difference of values of the integrand between the two ends of ∂V. This identity is equivalent to the following formula:

$$\begin{aligned} -\delta H_V(\varphi, \pi) &= \int_V \left(\dot{\pi} \delta \varphi - \dot{\varphi} \delta \pi \right) \, d\xi + \left[\pi^1 \delta \varphi \right]_{\partial V} \\ &= \int_V \left((\mathscr{L}_X \pi) \delta \varphi - (\mathscr{L}_X \varphi) \delta \pi \right) + \left[\pi^1 \delta \varphi \right]_{\partial V} , \end{aligned} \tag{2.44}$$

[2] Actually, $\pi d\xi$ is a pull-back of a differential odd-form $\pi^\mu \partial_\mu \rfloor d\xi^0 \wedge d\xi^1$ to the surface $\tau = \text{const.}$ This proves that it does not depend upon the choice of the coordinate x^0, but only upon the choice of the Cauchy surface Σ. The structure of the canonical field momentum π^μ is discussed in the next chapter.

where

$$H_V(\varphi, \pi) := \int_V (\pi\dot{\varphi} - \mathscr{L}) \ . \tag{2.45}$$

The procedure used applies to an arbitrary Lagrangian. Its geometric context, together with the structure of the momentum π^μ will be discussed thoroughly in the next chapter. Here, we perform calculations for the specific Lagrangian (2.37) and obtain:

$$\pi = \pi^0 = \frac{\partial \mathscr{L}}{\partial f_0} = \partial_\tau f + \xi \frac{\partial f}{\partial \xi} = \dot{\varphi} + \xi \partial_\xi \varphi \ , \tag{2.46}$$

$$\pi^1 = \frac{\partial \mathscr{L}}{\partial f_1} = \xi \left(\partial_\tau f + \xi \frac{\partial f}{\partial \xi} \right) - \frac{\partial f}{\partial \xi} = \xi \dot{\varphi} - (1 - \xi^2) \partial_\xi \varphi \ . \tag{2.47}$$

These equations imply, according to (2.41), the following field dynamics:

$$\dot{\varphi} = \pi - \xi \partial_\xi \varphi \ , \tag{2.48}$$

$$\dot{\pi} = \partial_\xi \left(\xi \dot{\varphi} - (1 - \xi^2) \partial_\xi \varphi \right) = \partial_\xi (\xi \pi) - \partial_\xi^2 \varphi \ . \tag{2.49}$$

Using (2.48), the Hamiltonian (2.45) may by written explicitely in terms of Cauchy data:

$$\begin{aligned} H_V(\varphi, \pi) &:= \int_V \{ \pi(\pi - \xi \partial_\xi \varphi) - \mathscr{L} \} \, d\xi \\ &= \frac{1}{2} \int_V \left\{ \pi^2 - 2\pi \xi \partial_\xi \varphi + (\partial_\xi \varphi)^2 \right\} d\xi \\ &= \frac{1}{2} \int_V \left\{ (\pi - \xi \partial_\xi \varphi)^2 + (1 - \xi^2)(\partial_\xi \varphi)^2 \right\} d\xi \ . \end{aligned} \tag{2.50}$$

The main lesson stemming from the above formulae is the following: Consider, first, a domain V that lies strictly inside the interior of the cone, *i.e.*, $V = [a, b]$, with $-1 < a, b < 1$. Then $\mathbb{R} \times \partial V$ is timelike and the situation is similar to the one described in Sect. 2.2: boundary conditions on ∂V have to be added and the mixed Cauchy- (on V) and boundary- (on ∂V) problem is well posed. Restricting the class of admissible functions to those fulfilling Dirichlet condition (2.16), we obtain (2.17) and, therefore, the boundary term in (2.44) vanishes. This implies the Hamiltonian form of the field evolution within the cone. A treatment similar to the one used in Sect. 2.2 is also applicable in the space of functions fulfilling Neumann conditions on ∂V.

The situation changes drastically if we pass to sections of the light cone, setting $V = [-1, 1]$. Having chosen Cauchy data (φ, π) on V at a given instant of time, say τ_0, we still have the freedom to chose boundary data on ∂V, *i.e.*, the values $\varphi(\tau, -1)$ and $\varphi(\tau, 1)$, but *only for* $\tau \leq \tau_0$. Indeed, with this whole set of Cauchy and boundary data, the scalar field f is uniquely determined within the entire light cone $\mathscr{C}^-$. Consequently, the values $\varphi(\tau, -1)$ and $\varphi(\tau, 1)$ for $\tau > \tau_0$ are uniquely determined by the Cauchy data and *cannot be chosen freely*. Even if we take trivial boundary data in the past, the boundary term in (2.43) might cease to vanish at an instant of time $\tau_0 + \epsilon$, and things can be arranged so that this happens arbitrarily close to τ_0. At a first glance there is no way to obtain a Hamiltonian evolution of Cauchy data in this case.

The remedy for these difficulties consists in treating the data on the boundary of the light cone not as *boundary* data, but as a further piece of Cauchy data. For this purpose, we extend the parameterization Equations (2.33)-(2.34) to $|\xi| > 1$ setting:

$$t = -x := -e^{-\tau+\xi-1} \qquad \text{for } \xi > 1 , \tag{2.51}$$

$$t = x \quad := -e^{-(\tau+\xi-1)} \qquad \text{for } \xi < -1 , \tag{2.52}$$

and we consider the data (φ, π) on the entire surface $\Sigma = \{\xi \in \mathbb{R}^1\}$. Within the interior of the light cone, *i.e.*, for $|\xi| < 1$, the dynamics is governed by the Hamiltonian (2.50):

$$H_{[-1,1]}(\varphi, \pi) = \frac{1}{2} \int_{-1}^{1} \{(\pi - \xi \partial_\xi \varphi)^2 + (1 - \xi^2)(\partial_\xi \varphi)^2\} \, d\xi , \tag{2.53}$$

which, as we have already seen, satisfies the correct Hamiltonian equation for the wave equation, modulo the boundary term in (2.44) which will be taken care of by the considerations that follow. Outside of the interval $\xi \in [-1, 1]$, the dynamics reduces to translations tangent to the hypersurface on which the data are given, as discussed in the previous section. The only difference here is that the relevant part of the phase spaces "lives" on a null rather than a spacelike hypersurface, which plays no role in the considerations of Sect. 2.3. More precisely, equation (2.51) implies that we have $X = \partial_\tau = \partial_\xi$ for $\xi < -1$, whereas (2.52) implies: $X = \partial_\tau = -\partial_\xi$ for $\xi > 1$. Consequently, we have:

$$\mathscr{L}_X \varphi = -\partial_\xi \varphi , \qquad \mathscr{L}_X \pi = -\partial_\xi \pi \qquad \text{for } \xi > 1 , \tag{2.54}$$

and

$$\mathscr{L}_X \varphi = \partial_\xi \varphi , \qquad \mathscr{L}_X \pi = \partial_\xi \pi , \qquad \text{for } \xi < -1 , \tag{2.55}$$

which are special cases of the formula (2.29). According to the standard procedure, described in the next chapter (see also footnote 2, page 17), the momentum π on Σ is taken as the pull-back to the Cauchy surface, of the differential (odd) form $\pi^\mu \partial_\mu \rfloor d\xi^0 \wedge d\xi^1$, with π^μ is given by formula (2.39). According to (2.30), the contribution to the Hamiltonian of the field contained in the region $[1, \infty)$ equals

$$H_{[1,\infty)}(\varphi, \pi) = \int_1^\infty (-\pi \partial_\xi \varphi) \, d^3x , \tag{2.56}$$

whereas the corresponding contribution from the region $[1, \infty)$ is equal to

$$H_{(-\infty,-1]}(\varphi, \pi) = \int_{-\infty}^{-1} (+\pi \partial_\xi \varphi) \, d^3x . \tag{2.57}$$

Let us prove that the functional H, equal to the sum of these contributions,

$$H := H_{(-\infty,-1]} + H_{[-1,1]} + H_{[1,\infty)} , \qquad (2.58)$$

satisfies the equation (2.2) defining a Hamiltonian for the joint dynamical system, given by (2.55) for $\xi < -1$, by (2.54) for $\xi > 1$ and by Equations (2.48)-(2.49) for $-1 < \xi < 1$. Indeed, a variation of H gives us the sum of two formulae of the type (2.31), for $\xi < -1$ and $\xi > 1$ respectively, together with formula (2.44) for $-1 < \xi < 1$. This means that we have:

$$\begin{aligned} -\delta H(\varphi, \pi) = & \int_{\Sigma} (\mathscr{L}_X \pi \delta\varphi - \mathscr{L}_X \varphi \delta\pi) \, d\xi \\ & + \left[\pi^1 \delta\varphi\right]_{-\infty}^{-1} + \left[\pi^1 \delta\varphi\right]_{-1}^{1} + \left[\pi^1 \delta\varphi\right]_{1}^{\infty} , \end{aligned} \qquad (2.59)$$

with appropriate values for $(\mathscr{L}_X \varphi, \mathscr{L}_X \pi)$ in the respective regions of Σ. But the intermediate boundary terms at $\xi = -1$ and $\xi = 1$ cancel because of the continuity of ξ and π^1. Assuming sufficiently strong fall-of conditions for the Cauchy data (*e.g.*, assuming that they are compactly supported on Σ) we also obtain a cancelation of the boundary terms at both infinities. What remains is the desired Hamiltonian formula for the total dynamics on Σ:

$$-\delta H(\varphi, \pi) = \int_{\Sigma} (\mathscr{L}_X \pi \delta\varphi - \mathscr{L}_X \varphi \delta\pi) \, d\xi . \qquad (2.60)$$

The continuity of π^1, which if fundamental for the cancelation of the intermediate boundary terms in (2.59), may be roughly explained as follows: π^1 is a component of the vector density π^μ, corresponding to the family of hypersurfaces $\xi^1 = \xi = \text{const.}$ This vector density is defined, and continuous, on the entire space-time (which in our case is the light cone $\mathscr{C}^-$). At boundary points $\xi = -1$ and $\xi = 1$ there is no jump in the field of tangents to the surfaces $\xi = \text{const.}$ and, therefore, π^1 is continuous as well. We note that there are various delicate issues concerning the *exterior* orientation of the hypersurfaces involved, which are discussed in the next chapter and in Appendix A.

2.5 The Trautman–Bondi energy for the scalar field

Formula (2.60) enables us to describe the dynamics of a massless scalar field within the light cone in terms of a Hamiltonian dynamical system in an abstract space $\mathbb{R}^2$, parameterized by the "generalized time parameter" τ and the "generalized space parameter" ξ. Cauchy data are given on the Cauchy surfaces $\Sigma = \{\xi^0 = \tau = \text{const.}\}$. The phase space $\mathscr{P} = \{(\varphi, \pi)\}$ is defined as the collection of compactly supported fields on Σ. The function φ is supposed to be continuous and piecewise smooth. On the other hand, the momentum

$\pi = \pi^0$ might fail to be continuous at boundary points $\xi = -1$ and $\xi = 1$, because the Cauchy surfaces $\Sigma = \{\xi^0 = \tau = \text{const}\}$ "change direction" in a non-continuous way there. Consequently, we assume that π is piecewise continuous in the three regions of Σ separately. Moreover, $\pi^0 \equiv \pi$ has to fulfill the following constraint:

$$\pi = -\dot{\varphi} = \partial_\xi \varphi \quad \text{for } \xi < -1 \, , \tag{2.61}$$

$$\pi = \dot{\varphi} \quad = -\partial_\xi \varphi \quad \text{for } \xi > 1 \, . \tag{2.62}$$

This can be seen from Equation (2.47) and from the fact that π^1 coincides with $\pi = \pi^0$ for $|\xi| > 1$; this last property holds because the hypersurfaces $\{\xi^1 = \text{const}\}$ coincide with the hypersurfaces $\{\xi^0 = \text{const}\}$ there.

The above constraints imply the following formulae for the Hamiltonian on the light cone:

$$H_{(-\infty,-1]} = \int_{-\infty}^{-1} (\partial_\xi \varphi)^2 \, d^3x \, , \tag{2.63}$$

and

$$H_{[1,\infty)} = \int_{1}^{\infty} (\partial_\xi \varphi)^2 \, d^3x \, . \tag{2.64}$$

(The reader is referred to equations (4.38)-(4.40) for an explicit calculation of the associated variational formulae in a similar context.)

The evolution with respect to the field $X = \partial_\tau$ is determined by the Hamiltonian formula (2.60), where the Hamiltonian is defined by (2.58). The Hamiltonian system obtained this way is autonomous, because the Hamiltonian does not depend explicitly on time. Hence, it is conserved during the evolution. But formulae (2.63) and (2.64) prove that $H_{(-\infty,-1]}$ and $H_{[1,\infty)}$ are monotonically increasing functions of time. Indeed, their values are equal to the integral of a non-negative function $(\partial_\xi \varphi)^2$ over a portion of the boundary $\partial\mathscr{C}^-$ of the cone which grows when time increases. This implies that $H_{[-1,1]}$ must be a monotonically decreasing function of time. We see that, due to radiation, the energy is being transferred from the "Cauchy zone": $[-1,1]$, to the "radiation zone": $[-\infty,-1] \cup [1,\infty]$.

The real radiation problem is obtained when the boundary $\partial\mathscr{C}^-$ is moved to infinity. By analogy with similar constructions done in general relativity, in such a case the amount of energy $H_{[-1,1]}$ contained in the Cauchy zone will be called the Trautman-Bondi energy of the scalar field. As we shall see in the sequel, the properties of $H_{[-1,1]}$ are analogous to the decreasing properties of the Trautman-Bondi mass in general relativity, irrespective of the limiting transition mentioned.

It should be pointed out that in the simple model of this section we have considered field configurations defined globally on $\mathscr{C}^-$. For various reasons, discussed below, it is useful to consider situations in which this is not the case. This introduces some supplementary complications, which are taken care of in the remainder of this monograph.

3. Hamiltonian flows for geometric field theories

Some twenty years ago Kijowski and Tulczyjew [95] have laid the ground for a geometric formulation of Hamiltonian analysis of the dynamics of field theories. In that reference the dynamics of fields was that corresponding to motions of a hypersurface Σ along the flow of a vector field X on a manifold M. The vector field was further assumed to be transverse to Σ. There are, however, several situations, where the transversality of X to Σ is not a natural restriction: this happens *e.g.* when considering spacelike translations in Minkowski space-time. Further, in some situations it might be convenient to consider motions of hypersurfaces which do not arise from the flow of a vector field; we shall indeed encounter such motions in Sect. 4.5 below. The purpose of this chapter is to generalize the framework of Kijowski and Tulczyjew to allow for such situations.

3.1 The framework

Consider a theory of sections $f = (f^A)$ of a fiber bundle F over an $(n+1)$-dimensional manifold M (a generalized space-time[1]):

$$\begin{array}{c} F \\ \Big\downarrow {\scriptstyle pr_{F \to M}} \\ M \end{array} \tag{3.1}$$

We shall throughout use the notation $pr_{Y \to X}$ for the bundle projection map of a bundle Y over a manifold X. Let Σ be an n-dimensional piecewise smooth manifold and let $i : \Sigma \to M$ be an embedding of Σ in M; thus $i(\Sigma)$ is a topological, piecewise smooth, hypersurface in M. The hypersurface $i(\Sigma)$

[1] The word "space-time" should not be understood as implying that there is some Lorentzian metric on M involved: we only use this term to denote the arena used to describe the dynamics of the fields under consideration. The adjective "generalized" refers to the following: 1) we are considering arbitrary dimensions n, not necessarily $n = 3$; 2) in some degenerate situations in the gravitational field case, the set M does not necessarily coincide with the actual space-times associated to the relevant initial data sets, see Sect. 5.7 for details.

should be thought of as a hypersurface on which appropriate data (*e.g.*, initial data) for the fields under consideration are given. Using i, we can pull-back the bundle F to a bundle Φ over Σ:

$$\begin{array}{ccc} \Phi & \xrightarrow{i!} & F \\ \Big\downarrow{\scriptstyle pr_{\Phi\to\Sigma}} & & \Big\downarrow{\scriptstyle pr_{F\to M}} \\ \Sigma & \xrightarrow{i} & M \end{array} \tag{3.2}$$

The map $i!$ above [2] preserves the bundle structure, and is an isomorphism on fibers. Thus, sections f of F defined over $i(\Sigma)$ are in one-to-one correspondence with sections $\phi = (\phi^A)$ of Φ over Σ:

$$\phi(\xi) = (i!_\xi)^{-1}(f(i(\xi))) , \tag{3.3}$$

where $i!_\xi$ denotes the restriction of $i!$ to a fiber of Φ over ξ. We emphasize that it will be essential for the Hamiltonian description of the dynamics at $\mathscr{I}^+$ that Σ and/or $i(\Sigma)$ is not smooth but piecewise smooth only. Further, in some of our applications, the piecewise smooth manifold F will not be a bundle; rather it will be the union of bundles defined over the smooth components of Σ. By an abuse of terminology we shall continue to call such objects bundles, hoping that it will not lead to confusions.

Let us start by recalling the construction of Kijowski and Tulczyjew [95] of some objects that naturally occur in geometric field theories. For any point $\xi \in \Sigma$ one defines a phase space at ξ:

$$\Pi_\xi \equiv T^*\Phi_\xi \otimes \tilde{\Lambda}^n T^*_\xi \Sigma ,$$

where Φ_ξ is the corresponding fiber of Φ, and $\tilde{\Lambda}^n T^*\Phi_\xi$ denotes the bundle of odd n-forms[3] over Φ_ξ. The collection of all spaces Π_ξ over $\xi \in \Sigma$ will be denoted by Π. It is again a fiber bundle over Σ. Actually, we have

[2] The bundle Φ is called the *induced* bundle and denoted by F' in [116, p. 391], while the map $i!$ is denoted by $\tilde{\imath}$ there. Dieudonné [53, Chap. XVI, Sect. 19, p. 127] denotes Φ by i^*F and calls it the inverse image of F.

[3] If M is orientable, $\tilde{\Lambda}^\ell$ can be identified with Λ^ℓ, when an oriented atlas is chosen. However, it should be remembered that *regardless* of the question of orientability of M the integral of an odd form is defined *differently* from that of a normal form: the usual ℓ-forms are integrated on ℓ-dimensional submanifolds *with interior orientation*, while odd ℓ-forms are integrated on ℓ-dimensional submanifolds *with exterior orientation*. This distinction is irrelevant in our analysis of the scalar field, Chap. 4 below, once the (usual) choices are made that the exterior orientation of an achronal hypersurface is determined by a transverse vector which is timelike and future directed; then the obvious exterior orientation of a coordinate ball included in a hypersurface is chosen. However, a careful bookkeeping of the exterior orientation of the sets considered becomes important in order to get correct signs of various integrals in the gravitational case, Chap. 5, where natural coordinate systems near $\mathscr{I}^+$ have an orientation which is opposite to the natural orientation of the space-time. For the convenience of the reader we give an overview of odd n-forms in Appendix A.

$$\Pi \equiv V^*\Phi \otimes \tilde{\Lambda}^n T^* \Sigma \,,$$

where $V^*\Phi$ denotes the bundle of vertical covectors[4] on Φ (elements dual to vertical vectors over Φ). Elements of Π_ξ are "scalar-density-valued" covectors on Φ_ξ. Since the factor $\tilde{\Lambda}^n T^*_\xi \Sigma$ in the tensor product is constant over each fiber Π_ξ, it may be put "outside of the bracket". In this way the standard construction (*cf.*, *e.g.*, [2, 95]) of a canonical one-form and a canonical (symplectic) two-form in a cotangent bundle may be performed (at each ξ separately) almost without any change:

$$\Theta_\xi(\pi) := pr^*_{\Pi\to\Phi}\pi \,,$$

where by $pr_{\Pi\to\Phi}$ we denote the canonical projection from Π to Φ. Moreover,

$$\Omega_\xi := d\Theta_\xi \,,$$

where d denotes the exterior derivative operator on the manifold Π_ξ (ξ being fixed). Because of the geometric nature of Π, the forms obtained this way are no longer "scalar-valued" but rather "scalar-density-valued".

If (ξ^i) are coordinates on Σ and (ϕ^A, ξ^i) are coordinates on Φ, then every element of Π may be represented by a pair: a n-form

$$\pi = \pi_A d\phi^A \otimes (d\xi^1 \wedge \cdots \wedge d\xi^n) \,,$$

and the orientation carried by this coordinate system (see Appendix A for a detailed discussion). The coefficients π_A may be used to define the coordinate system (π_A, ϕ^A, ξ^i) in Π. In terms of these coordinates we have:

$$\Theta_\xi = \pi_A d\phi^A \otimes (d\xi^1 \wedge \cdots \wedge d\xi^n) \,,$$

and

$$\Omega_\xi = (d\pi_A \wedge d\phi^A) \otimes (d\xi^1 \wedge \cdots \wedge d\xi^n) \,.$$

"Integrating" the forms Ω_ξ over Σ gives us, loosely speaking, a "symplectic form" Ω_Σ in the space of sections of the phase bundle Π. (In some cases Ω_Σ will be indeed a symplectic form, when appropriate function spaces have been specified. We will avoid the term "symplectic" since we do not wish to address those issues.) Loosely speaking again, "vectors tangent to the space of sections" may be represented as fields of vertical vectors (*i.e.*, vectors tangent to fibers):

$$\Sigma \ni \xi \to X(\xi) \in T\Pi_\xi \,.$$

If X_1 and X_2 are two such fields, attached at the same section of Π (*i.e.*, $X_1(\xi)$ and $X_2(\xi)$ are attached at the same point of Π_ξ), then we put:

[4] An element of $V^*\Phi$ may be represented as an equivalence class of elements of $T^*\Phi$: two elements $\alpha, \beta \in T^*_\phi\Phi$ are equivalent if $\alpha(v) = \beta(v)$ for any vertical vector $v \in V_\phi\Phi \subset T_\phi\Phi$.

$$\Omega_\Sigma(X_1, X_2) := \int_\Sigma \Omega_\xi(X_1(\xi), X_2(\xi)) \ .$$

If the "vectors" X_a, $a = 1, 2$, have the following coordinate representation:

$$X_a(\xi) = (\delta_a \pi_A(\xi), \delta_a \phi^B(\xi)) \ ,$$

then the coordinate expression for Ω_Σ reads:

$$\Omega_\Sigma((\delta_1 \pi_A, \delta_1 \phi^B), (\delta_2 \pi_A, \delta_2 \phi^B)) =$$
$$\int_\Sigma \{(\delta_1 \pi_A(\xi))(\delta_2 \phi^A(\xi)) - (\delta_2 \pi_A(\xi))(\delta_1 \phi^A(\xi))\} \, d^n\xi \ , \quad (3.4)$$

whenever the integrand above is in $L^1(\Sigma)$. We stress that the above formula is invariant with respect to changes of coordinates (ϕ^A, ξ^i) in Φ. In particular, the coordinates π_A behave like scalar densities with respect to changes of coordinates (ξ^i) on Σ. Hence, a Jacobian factor arising from the change of variables is absorbed by the corresponding factor in the transformation law for π_A.

In this section we are going to show that every motion of $i(\Sigma)$ in M leads to a class of dynamical systems on the space of appropriately behaved sections of Π. Moreover, those dynamical systems will be Hamiltonian with respect to the form Ω_Σ, in the sense described in Sect. 2.1.

Let, thus, I be an interval in $\mathbb{R}$, and let $\psi : I \times \Sigma \to M$ be a piecewise smooth map such that for each $\tau \in I$ the maps

$$\Sigma \ni \xi \to \psi_\tau(\xi) \equiv \psi(\tau, \xi) \in M$$

are homeomorphisms of Σ and $\psi_\tau(\Sigma)$, piecewise diffeomorphisms, with

$$\psi_0 = i \ .$$

Such a map ψ can be thought of as a motion of $i(\Sigma)$ in M. We emphasize, however, that the maps do not necessarily *move* $i(\Sigma)$ in M, for example, the ψ_τ's could be translations along $i(\Sigma)$. Further, we wish to stress that while the ψ_τ's are assumed to be bijective from Σ to $\psi_\tau(\Sigma)$, we do *not* assume that ψ is; translations along $i(\Sigma)$ give an example of a ψ which will not be bijective. Given a map ψ, to define a dynamical system on the space of sections of the phase bundle Π we need a bundle map $\Psi : I \times \Phi \to F$ such that the diagram

$$\begin{array}{ccc}
I \times \Phi & \xrightarrow{\Psi} & F \\
{\scriptstyle id_\mathbb{R} \times pr_{\Phi \to \Sigma}} \downarrow & & \downarrow {\scriptstyle pr_{F \to M}} \\
I \xleftarrow{} I \times \Sigma & \xrightarrow{\psi} & M
\end{array} \quad (3.5)$$

commutes. For $\phi \in \Phi$ we set

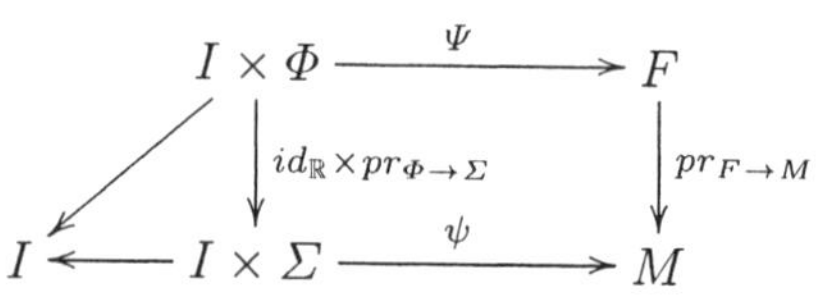

$$\Psi_\tau(\phi) = \Psi(\tau, \phi) \,. \tag{3.6}$$

In (3.5) the two leftermost arrows are projections on the I factor of the Cartesian product. We shall further require that for all $\tau \in I$ the map $\Psi(\tau, \cdot)$ is an isomorphism from the fibers of Φ to those of F.

We emphasize that Ψ is arbitrary up to the requirement of commutativity of the diagram (3.5), and that different choices of Ψ will lead to different dynamical systems on the space of sections of the phase bundle Π. Indeed, Ψ plays a role similar to that of a reference frame in mechanics. It enables us to compare configurations of the physical system at different times. Whether or not the particle moves *is not* an absolute statement but depends upon a choice of a reference frame. The Hamiltonian description of the particle dynamics is only possible *with respect to a chosen reference frame.* The same dynamics, described with respect to different reference frames, leads to different Hamiltonian systems. Similarly, the Hamiltonian description of the field dynamics is possible only *with respect to a given reference system* Ψ.

We further stress that there are several ways of prescribing Ψ for a given ψ. As an example, consider the following three useful ways of doing this:

- Suppose that ψ is obtained from a flow of a vector field X defined on M, and that F is a bundle of *geometric objects* in the sense that the action of diffeomorphisms on M lifts canonically to an action of diffeomorphisms of F. One can then use the canonical lift of ψ to obtain Ψ. For example, if F is a tensor bundle over M, then the so obtained action of Ψ on a tensor field f consists, essentially, of the Lie dragging f along the orbit of X until the orbit meets $i(\Sigma)$.
- When F is a tensor bundle, a different, often used (*cf., e.g.,* [80]), construction of Φ proceeds as follows: Let g be a metric on M (which might, but does not have to, be part of the set of the fields f under consideration), and suppose further that the hypersurfaces

$$S_\tau \equiv \psi_\tau(\Sigma) \tag{3.7}$$

are spacelike. In that case one can use the metric g to decompose each tangent space $T_{\psi(\tau,\xi)}M$ into a part which lies in the image of $\psi(\tau,\cdot)_*$ and its orthogonal complement. The first part of this decomposition is naturally diffeomorphic to $T_\xi\Sigma$, while the latter is diffeomorphic to $\mathbb{R}$. There is a corresponding decomposition of higher tensor bundles over $\psi(I \times \Sigma) \subset M$ induced by this decomposition of $T_{\psi(\tau,\xi)}M$. For example, a symmetric two-covariant tensor field on M is decomposed in this way into a symmetric tensor field on Σ, a covector field ("the shift") on Σ, and a scalar field ("the lapse") on Σ. This gives rise to a map Ψ after an identification of Φ with a finite sum of tensor bundles over Σ.
- When F is a vector or affine bundle equipped with a connection (which might, but does not have to, be part of the set of the fields f under consideration), then Ψ can be constructed as follows: For $\xi \in \Sigma$ let γ_ξ denote the path

$$I \ni \tau \to \gamma_\xi(\tau) \equiv \psi(\tau, \xi) \,. \tag{3.8}$$

For ϕ belonging to the fiber Φ_ξ of Φ over $\xi \in \Sigma$ one defines $\Psi(\tau, \phi) \in F$ by parallel transporting $\Psi_{(0,\xi)}(0, \phi) = i!_\xi(\phi)$ from $i(\xi) = \psi(0, \xi)$ to $\psi(\tau, \xi)$ along γ_ξ. Here $\Psi_{(\tau,\xi)}$ denotes the restriction of Ψ_τ to a fiber of Φ over ξ, while $i!_\xi$ is that of $i!$ to the same fiber.

As an illustration, consider a theory of a tensor field in Minkowski space-time $M = \mathbb{R}^{1,n}$, and let ψ be obtained from the flow of a vector field X defined on M. There is a natural choice for the metric g on M — the defining one, and there is also the corresponding flat connection. Each of the three prescriptions above will define a different reference system on the space of sections of Π except in special cases, such as that of X — a translational Killing vector field on M and $i(\Sigma)$ — a hypersurface of standard Minkowski time in M.

It is useful to describe exhaustively the freedom in defining the maps Ψ satisfying our requirements. We note, first, that Ψ can be used to define parallel transport in F along the paths γ_ξ defined by equation (3.8): a section of F over γ_ξ is horizontal if its counterpart over $I \times \{\xi\}$ is τ-independent. An equivalent way of defining this is to introduce a *covariant derivative along the vector X tangent to γ_ξ*. At each point $\psi(\tau, \xi)$ the vector X is given by the equation

$$X \equiv \psi_* \frac{\partial}{\partial \tau} \,. \tag{3.9}$$

If f is a section of F over γ_ξ, then $\Psi^{-1}_{(\tau,\xi)} f(\psi(\tau, \xi)) \in \{0\} \times \Phi_\xi$ and, therefore, $\frac{d}{d\tau}\left(\Psi^{-1}_{(\tau,\xi)} f(\psi(\tau, \xi))\right)$ is an element of $T\Phi_\xi$; one then sets

$$\mathcal{D}_X f(\psi(\tau, \xi)) = \Psi_* \frac{d(\Psi^{-1}_{(\tau,\xi)} f(\psi(\tau, \xi)))}{d\tau} \in TF_{\psi(\tau,\xi)} \,. \tag{3.10}$$

Conversely, given a connection defined on sections over γ_ξ, for ϕ in the fiber Φ_ξ of Φ over ξ we define $\Psi(\tau, \phi) \in F$ by parallel transporting $\Psi_{(0,\xi)}(\phi)$ from $i(\xi) = \psi(0, \xi)$ to $\psi(\tau, \xi)$ along γ_ξ. This shows that defining Ψ is equivalent to prescribing a connection *along sections over the γ_ξ's*, together with $\Psi_0 \equiv \Psi|_{\tau=0}$. (There is a natural choice $\Psi_0 = i!$, but it might be convenient in some cases to use other Ψ_0's.) This connection can, but does not need to, arise from a connection on F.

For each τ equation (3.9) defines a vector field X_τ defined on $\psi_\tau(\Sigma)$ because, for all $\tau \in I$, ψ_τ is a diffeomorphism between Σ and $\psi_\tau(\Sigma)$. We emphasize that X is not a space-time vector field in general, which can be seen by considering a motion in which Σ is moved first forwards then backwards in time, so that the image of Σ at some later time crosses that at an earlier time.

For further purposes, let us recall that a connection on sections of F over γ_ξ induces naturally a corresponding connection on V^*F. A necessary intermediate object is a connection in VF, and the corresponding connection on V^*F is obtained as a dual to the latter. More precisely, if r is a section of V^*F over γ_ξ, then for any section δf of VF, $r(\delta f)$ is a function on γ_ξ. The quantity $\mathcal{D}_X r$ is then uniquely defined by the requirement that

$$\frac{d(r(\delta f))}{d\tau} = (\mathcal{D}_X r)(\delta f) + r(\mathcal{D}_X \delta f) , \tag{3.11}$$

for all δf.

The construction of the connection in VF from a connection in F is especially simple in the case when F is a vector bundle. In this case elements of VF (vectors tangent to a vector space) may be identified with elements of F (elements of the space). Due to this identification a connection on F implies immediately the corresponding connection on VF.

In the general case any section of VF may be represented as a derivative of a one-parameter family $f(\lambda)$ of sections of F with respect to λ. As in (2.8), throughout the paper we use δ to denote the derivative with respect to this external parameter. We define a connection in VF by the following formula:

$$\mathcal{D}_X(\delta f^A) := \delta(\mathcal{D}_X f^A) .$$

Recall, now, that a connection on a bundle F_{γ_ξ} defined above a path γ_ξ with tangent vector X can be defined by prescribing a vertical vector field Γ on F_{γ_ξ}, equal to the difference between the vector field $\Psi_* \partial/\partial\tau$ and the horizontal lift of X. In local coordinates we have

$$(\mathcal{D}_X f^A)(\psi(\tau,\xi)) = \frac{d(f^A(\psi(\tau,\xi)))}{d\tau} + \Gamma^A(f,\psi(\tau,\xi)) . \tag{3.12}$$

This implies

$$(\mathcal{D}_X \delta f^A)(\psi(\tau,\xi)) = \frac{d(\delta f^A(\psi(\tau,\xi)))}{d\tau} + \Gamma^A_B(f,\psi(\tau,\xi))\delta f^B , \tag{3.13}$$

where $\Gamma^A_B = \partial\Gamma^A/\partial f^B$. Consequently,

$$\mathcal{D}_X r_A(\psi(\tau,\xi)) = \frac{d(r_A(\psi(\tau,\xi)))}{d\tau} - \Gamma^B_A(f,\psi(\tau,\xi)) r_B . \tag{3.14}$$

In the linear case the vector Γ depends linearly upon f. Hence, there is a choice of coordinates such that

$$\Gamma^A(f,x) = \Gamma^A_B(x) f^B ,$$

so that the connection coefficients Γ^A_B are constant along the fibres of F.

3.2 Hamiltonian dynamics for Lagrangian theories

Now, we are going to show how to translate the field dynamics, described it terms of partial differential equations on M, into a language of Hamiltonian dynamics in the phase space of sections of Π. For this purpose we fix a field reference system Ψ. Let, then, f be a section of F. For $\tau \in I$ the equation

$$\phi(\tau,\xi) = \Psi^{-1}_{(\tau,\xi)} f(\psi(\tau,\xi)) \tag{3.15}$$

defines the one-parameter family of sections of Φ (a curve in an infinite dimensional configuration space). We shall write $\phi(\tau)$ for $\phi(\tau,\cdot)$ when no ambiguities are likely to occur.

Because canonical momenta are described by sections of $\Pi = V^*\Phi \otimes \tilde{\Lambda}^n T^*\Sigma$, we need also a counterpart of this space over space-time M. This is provided by the *space-time phase bundle* P defined as

$$P \equiv V^*F \otimes \tilde{\Lambda}^n T^*M \ .$$

As before, we use the symbol $\tilde{\Lambda}^\ell T^*M$ to denote the bundle of odd ℓ-forms. In local coordinates odd forms are represented by normal forms, up to an overall sign related to the orientation of the coordinate system. It follows that in local coordinates every element $p \in P$ can be written as

$$p = p_A{}^\mu df^A \otimes (\partial_\mu \rfloor dx^1 \wedge \ldots dx^{n+1}) \ , \tag{3.16}$$

where $\rfloor$ denotes contraction. Hence, sections of P can be locally represented by the collection of functions $(p_A{}^\mu, f^A)$ defined on M.

To understand how sections of this bundle arise in field theory, suppose that the field dynamics for sections f is derived from a first order Lagrange function. More precisely, let J^1F denote the bundle of first jets of sections of F (*cf.*, *e.g.*, [110] or [98, Appendix 2]), and let $\mathbf{L}$ be a bundle map from J^1F to $\tilde{\Lambda}^{n+1}T^*M$; in local coordinates we can write

$$\mathbf{L} = L(f^A, f^A{}_\mu, x^\mu) dx^1 \wedge \ldots \wedge dx^{n+1} \ , \tag{3.17}$$

with $f^A{}_\mu \equiv \frac{\partial f^A}{\partial x^\mu}$. L defined in (3.17) is actually a density and not a function, since it "picks up" a Jacobian factor under changes of coordinates consistent with the orientation of M. The variational equations derived from L can be written in the form

$$p_A{}^\mu = \frac{\partial L}{\partial f^A{}_\mu} \ , \tag{3.18}$$

$$\partial_\mu p_A{}^\mu = \frac{\partial L}{\partial f^A} \ . \tag{3.19}$$

To any section f of F we can assign a section of P using equation (3.18). Then f will satisfy the Euler-Lagrange equations arising from the variational principle associated with L if and only if equations (3.18)–(3.19) hold. We will therefore represent dynamically admissible sections of F by *all* sections of P for which equations (3.18)–(3.19) hold.

We will derive now a short, convenient version of equations (3.18)–(3.19). Let $p(\lambda)$ be a one-parameter family of sections of P, with local coordinate representation $(p_A{}^\mu(\lambda), f^B(\lambda))$, satisfying[5] equations (3.18)–(3.19). We set $p_A{}^\mu \equiv p_A{}^\mu(0)$, and throughout the paper the symbol δ will denote $d/d\lambda|_{\lambda=0}$, *e.g.*

$$\begin{aligned} \delta L &\equiv \left.\frac{dL(j^1 f(\lambda), x^\mu)}{d\lambda}\right|_{\lambda=0} , \\ \delta f^A &\equiv \left.\frac{df(\lambda)}{d\lambda}\right|_{\lambda=0} , \end{aligned} \tag{3.20}$$

etc. The Euler–Lagrange equations (3.18)–(3.19) can be rewritten as

$$\begin{aligned} \delta L(f^A, f^A{}_\mu, x^\mu) &= \frac{\partial L}{\partial f^A}\delta f^A + \frac{\partial L}{\partial f^A{}_\mu}\delta f^A{}_\mu \\ &= (\partial_\mu p_A{}^\mu)\delta f^A + p_A{}^\mu \delta f^A{}_\mu \\ &= \partial_\mu (p_A{}^\mu \delta f^A) , \end{aligned} \tag{3.21}$$

or, equivalently,

$$\delta \mathbf{L} = \mathbf{d}(p(\delta f)) , \tag{3.22}$$

where $\mathbf{d}$ denotes the exterior derivative with respect to the common base manifold (M or $\mathbb{R} \times \Sigma$) of all the bundles involved.

Given a section p of P we are going to analyze it with respect to a given reference system Ψ. For this purpose we will transport it from M to $\mathbb{R} \times \Sigma$ *via* Ψ. In this way we obtain a one-parameter family of sections of Π

$$\pi(\tau) = \pi(\tau, \cdot) \tag{3.23}$$

given by

$$\pi(\tau, \xi) \equiv \left((\Psi_{(\tau,\xi)})^* \otimes \psi^*_{(\tau,\xi)} \right) p(\psi(\tau, \xi)) . \tag{3.24}$$

In local coordinates,

$$\pi(\tau) = (\pi_A(\tau), \phi^B(\tau)) , \tag{3.25}$$

where $\phi(\tau) = (\phi^B(\tau))$ has been defined in equation (3.15). We wish, now, to use this equation to define a dynamical system on the set of appropriate sections of Π. We define the phase space $\mathscr{P}_\Sigma$ as follows:

$$\begin{aligned} \mathscr{P}_\Sigma = \{ &\text{The space of sections of } \Pi \text{ which are obtained at } \tau = 0 \\ &\text{by the construction above from sections of } P \\ &\text{satisfying field Equations (3.18)–(3.19).} \} \end{aligned} \tag{3.26}$$

[5] For some purposes it might be convenient to assume that equations (3.18)–(3.19) are satisfied at $\lambda = 0$ only. We note that all our equations will still hold under this weaker condition. For esthetic reasons we prefer, however, to work only with solutions of the field equations.

The definition (3.26) needs to be complemented by the imposition of restrictions concerning the differentiability and the global behaviour of the sections of P involved, to ensure that the field equations are well defined and can be solved, and that the appropriate integrals converge. As we will see shortly, some further boundary conditions might have to be imposed. We note that the definition of $\mathscr{P}_\Sigma$ might lead to (Hamiltonian) constraints, when some sections of Π are not obtainable in the way prescribed in (3.26). Moreover, even if a section $\pi = \pi(0)$ *can* be obtained from a section p of P, the latter section may be highly non unique. Finally, $\pi(\tau)$ might not be in $\mathscr{P}_\Sigma$ even when $\pi(0)$ is. Nevertheless, in several cases of interest the curve

$$\forall \tau \in I \qquad T_\tau(\pi_A(0), \phi^B(0)) \equiv (\pi_A(\tau), \phi^B(\tau)) \tag{3.27}$$

turns out to be unique, is in $\mathscr{P}_\Sigma$, and equation (3.27) defines a dynamical system on $\mathscr{P}_\Sigma$. In several other cases of interest there will be a genuine ambiguity, and different choices of sections p of P which correspond to the same $\pi(0)$ will lead to different $\pi(\tau)$'s. The resulting non-uniqueness may sometimes be interpreted as a gauge transformation (*cf.*, *e.g.*, [2, 95]), so that equation (3.27) will instead define a dynamical system on the set of equivalence classes under gauge transformations of sections of F. We are not aware of any general approach to this question, which seems to require a case by case analysis.

To give the Hamiltonian description of the above dynamical system it is useful to keep track of the remaining information contained in p, encoded in the following object:

$$\boldsymbol{\pi} = (\Psi^* \otimes \psi^*)p \, . \tag{3.28}$$

In local coordinates (τ, ξ^a) on $I \times \Sigma$, and using the associated local orientation, the fields π and $\boldsymbol{\pi}$ can be represented by the forms

$$\pi = \pi_A d\phi^A \otimes d\xi^1 \wedge \ldots \wedge d\xi^n \, , \tag{3.29}$$

$$\boldsymbol{\pi} = \pi_A d\phi^A \otimes d\xi^1 \wedge \ldots \wedge d\xi^n + \pi_A{}^a d\phi^A \otimes \partial_a \rfloor d\tau \wedge d\xi^1 \wedge \ldots \wedge d\xi^n \, . \tag{3.30}$$

Now, we will show that for *any* solution p of field equations (3.18) and (3.19) and *any* reference system Ψ, formula (3.27) leads to a flow which is Hamiltonian with respect to the form Ω_Σ. This way we have translated the field dynamics over M into the language of the Hamiltonian flows. For this purpose we first "pull back" our Lagrangian $\mathbf{L}$ from M back to $\mathbb{R} \times \Sigma$:

$$\mathscr{L}(j^1\phi, \tau, \xi) := \psi^*(\mathbf{L}(j^1 f, \psi(\tau, \xi))) \, , \tag{3.31}$$

where f is any section of F such that ϕ is its pull-back, in the sense of equation (3.15), $\phi(\tau,\xi) = \Psi^{-1}_{(\tau,\xi)} f(\psi(\tau,\xi))$. This may again lead to (Lagrangian) constraints[6], if some sections ϕ of $\mathbb{R} \times \Phi$ do not admit the representation

[6] The term "Lagrangian constraints" may be understood in various ways: here we wish to stress that at those points (τ, ξ), at which ψ^* degenerates, the jets $j^1\phi$ of sections of Φ obtained by equation (3.15) will satisfy some linear relations.

(3.15). The analysis, which follows, works also in the presence of such constraints (*cf.*, *e.g.*, [94] for a more detailed description of the Hamiltonian formalism for constrained Lagrangian systems). In local coordinates, the above formula reads

$$\begin{aligned}\mathscr{L}_\tau d\tau \wedge d\xi^1 \wedge \ldots \wedge d\xi^n &\equiv \mathscr{L}(j^1\phi, \tau, \xi) \\ &= \mathscr{L}(\phi^A, \dot\phi^A, \phi^A{}_a, \tau, \xi^a) d\tau \wedge d\xi^1 \wedge \ldots \wedge d\xi^n \\ &= L(f^A, f^A{}_\mu, \psi^\mu(\tau, \xi^a))\, \psi^*(dx^1 \wedge \cdots \wedge dx^{n+1}) \\ &= L \frac{\partial(x^1, \ldots, x^{n+1})}{\partial(\tau, \xi^1, \ldots, \xi^n)} d\tau \wedge d\xi^1 \wedge \ldots \wedge d\xi^n \,. \qquad (3.32)\end{aligned}$$

Throughout the paper a dot over a symbol denotes a τ-derivative, while

$$\phi^A{}_a \equiv \frac{\partial f^A}{\partial \xi^a} \,.$$

We emphasize that there could be arbitrariness introduced in this definition, because there could be several fields f on space-time which correspond to a given ϕ on $\mathbb{R} \times \Sigma$. This, fortunately, does not lead to ambiguities in the definition of $\mathscr{L}$: if ψ is a diffeomorphism in a neighbourhood of (τ, ξ), then the first jet $j^1 f$ of f is determined uniquely by $j^1\phi$, so that $\mathscr{L}(\phi^A, \dot\phi^A, \phi^A{}_a, \tau, \xi^a)$ is uniquely defined. On the other hand, if ψ degenerates at (τ, ξ), then $\psi^*(dx^1 \wedge \cdots \wedge dx^{n+1})$ vanishes, so that $\mathscr{L}(\phi^A, \dot\phi^A, \phi^A{}_a, \tau, \xi^a)$ is zero, again uniquely defined.

We claim that the Hamiltonian governing the evolution (3.27) is given by the standard "Legendre transformation" formula from the above "pulled-back" Lagrangian:

$$\mathscr{H}_\Sigma(\pi, \tau) \equiv \mathscr{H}_\Sigma(\pi_A, \phi^B, \tau) \equiv \int_\Sigma (\pi_A(\tau)\dot\phi^A(\tau) - \mathscr{L}_\tau) d^n\xi \,. \qquad (3.33)$$

In order to see that $\mathscr{H}_\Sigma$ so defined generates indeed the appropriate evolution equation, equation (2.2), take a differentiable two-parameter family $\pi(\tau, \lambda)$ of sections of Π, obtained from any one-parameter family $p(\lambda)$ of sections of P, solutions of the field equations[5], *via* the transport (3.24). Denote $f(\lambda) := pr_{P \to F} p(\lambda)$. Taking the derivative of $\mathbf{L}(j^1(f(\lambda)))$ with respect to the parameter λ and applying ψ^* to both sides of equation (3.22), one obtains[7]

$$\delta\mathscr{L} = \delta(\psi^*\mathbf{L}) = \psi^*(\delta\mathbf{L}) = \psi^*(\mathbf{d}(p(\delta f))) = \mathbf{d}(\boldsymbol{\pi}(\delta\phi)) \qquad (3.34)$$

or, in terms of local coordinates:

[7] In some formulations of general relativity it might be convenient to allow maps ψ which are field dependent, in which case equation (3.34) would contain supplementary terms related to that dependence. We shall not use such an approach here.

$$\delta\mathscr{L}(j^1\phi(\tau,\lambda),\tau,\xi^a) = \frac{\partial(\pi_A(\tau)\delta\phi^A(\tau))}{\partial\tau} + \frac{\partial(\pi^a{}_A(\tau)\delta\phi^A(\tau))}{\partial\xi^a} . \quad (3.35)$$

It follows that

$$\delta(\pi_A\dot\phi^A - \mathscr{L}) = \dot\phi^A\delta\pi_A - \dot\pi_A\delta\phi^A - \partial_a(\pi_A{}^a\delta\phi^A) .$$

Inserting this equation into the variation of $\mathscr{H}_\Sigma$, one obtains[8]

$$-\delta\mathscr{H}_\Sigma(\pi_A,\phi^B,\tau) = \int_\Sigma \left\{\dot\pi_A\delta\phi^A - \dot\phi^A\delta\pi_A\right\} d^n\xi + \int_{\partial\Sigma} \pi_A{}^a\delta\phi^A d\sigma_a , \quad (3.36)$$

where $d\sigma_a = \partial_a \rfloor d\xi^1\wedge\ldots\wedge d\xi^n$ is a surface element on the boundary $\partial\Sigma$. Comparing with the definition of Ω_Σ, equation (3.4), we see that we obtain the desired formula, equation (2.2), *provided that the integral over $\partial\Sigma$ vanishes.* We emphasize that equation (3.36) holds when Σ has "creases" across which Σ is not differentiable, as long as $\delta\phi^A$ and the relevant components of π are continuous across them.

3.3 Space-time integrals

It is extremely useful to have the space-time equivalents of the above equations, so that everything can be calculated in terms of fields directly on the space-time M. First, the form Ω_Σ can be rewritten as

$$\begin{aligned}\Omega_\Sigma(\delta_1\pi,\delta_2\pi) &\equiv \Omega_\Sigma((\delta_1\pi_A,\delta_1\phi^B),(\delta_2\pi_A,\delta_2\phi^B))\\ &= \Omega_{i(\Sigma)}((\delta_1 p_A{}^\mu,\delta_1 f^B),(\delta_2 p_A{}^\mu,\delta_2 f^B))\\ &\equiv \Omega_{i(\Sigma)}(\delta_1 p,\delta_2 p)\\ &\equiv \int_{i(\Sigma)}\{(\delta_1 p)_A{}^\mu(\delta_2 f)^A - (\delta_2 p)_A{}^\mu(\delta_1 f)^A\}dS_\mu . \quad (3.37)\end{aligned}$$

Here the odd form dS_μ is defined as

$$\frac{\partial}{\partial x^\mu}\rfloor dx^1\wedge\cdots\wedge dx^{n+1} , \quad (3.38)$$

where $\rfloor$ denotes contraction. To obtain the space-time form of the Hamiltonian, we note the identity

[8] In equation (3.36) we have formally used the Stokes theorem to change a volume integral to a boundary integral; this is not allowed in general, as the Stokes theorem *per se* applies only to compact manifolds. We shall freely do such manipulations throughout this section without worrying whether or not this procedure is justified. When Σ is non compact the "integral over $\partial\Sigma$" is understood by a limiting procedure and the justification of formulae such as equation (3.36) requires a careful examination of the convergence of the integrals involved. This analysis will be performed when we will apply below the framework described here to the scalar field and to the gravitational field.

$$\begin{aligned}\mathscr{L} d\xi^1 \wedge \ldots \wedge d\xi^n &= \frac{\partial}{\partial \tau} \rfloor \mathscr{L} \\ &= \psi^*((\psi_* \frac{\partial}{\partial \tau}) \rfloor \mathbf{L}) \\ &= \psi^*(X \rfloor \mathbf{L}) \,, \end{aligned} \tag{3.39}$$

where X has been defined in equation (3.9). From the definition (3.24) and (3.29) of π, and from equation (3.10) one obtains

$$\mathscr{H}_\Sigma(\pi, \tau) = H(X, \psi_\tau(\Sigma), p) \,, \tag{3.40}$$

where

$$H(X, \mathscr{S}, p) \equiv \int_{\mathscr{S}} (p_A{}^\mu \mathcal{D}_X f^A - X^\mu L) dS_\mu \,. \tag{3.41}$$

If one wants to write a space-time equivalent of equation (3.36), one needs to define a space-time equivalent of $\dot{\pi}$. One way to implement this is to extend the derivative operator $\mathcal{D}_X$ to sections p of the bundle P. (This, on the other hand, can be implemented by choosing for each $\xi \in \Sigma$ a connection on sections of TM along the paths γ_ξ defined by equation (3.8).) It is important to realize that if one proceeds in this way, there is a new element of arbitrariness introduced, as there is no unique way of extending $\mathcal{D}_X$: while $\partial\pi/\partial\tau$ is uniquely defined once Ψ is given, there is no natural "$\partial/\partial\tau$ operation" on the remaining components of $\boldsymbol{\pi}$. This arbitrariness is, essentially, an arbitrariness of notation, as it will not affect the equations we consider here: in those equations terms with τ-derivatives other than those involving $\partial\pi/\partial\tau$ give no contribution once the integrals have been performed.

Let, then, $\mathcal{D}_X$ denote a covariant derivative operator on sections of P along the paths γ_ξ; it is convenient to restrict the freedom of choice of such operators by requiring

$$\psi_\tau^* \left((\mathcal{D}_X p)(f) + p(\mathcal{D}_X f) \right) = \frac{\partial \left(\psi_\tau^* \left(p(f) \right) \right)}{\partial \tau} \,. \tag{3.42}$$

We stress that while this is a natural restriction, it is only a matter of notational convenience, and that any other choice of connection can be used (leading, however, to formulae which do not look as natural as equation (3.43) below). We note that if ψ arises from the flow of a vector field X on the space-time M, then a natural choice of $\mathcal{D}_X$ is that arising from the operator (3.14) "acting on the A index of $p_A{}^\mu$" and the Lie derivative operator $\pounds_X$ "acting on the μ index". It can then be seen, using equation (3.11) and the fact that the Lie derivative commutes with the pull-back, that equation (3.42) holds with this choice of $\mathcal{D}_X p$.

With the constraint (3.42) on the action of $\mathcal{D}_X$ on sections of P, equation (3.36) takes the following space-time form:

$$-\delta H = \int_{\mathscr{S}} \left\{ (\mathcal{D}_X p_A{}^\mu) \delta f^A - (\mathcal{D}_X f^A) \delta p_A{}^\mu \right\} dS_\mu$$

$$+\int_{\partial\mathscr{S}} X^{[\mu}p_A{}^{\nu]}\delta f^A\,dS_{\mu\nu}$$

$$= \Omega_{\mathscr{S}}((\mathcal{D}_X p_A{}^\mu, \mathcal{D}_X f^A), (\delta p_A{}^\mu, \delta f^A)) + \int_{\partial\mathscr{S}} X^{[\mu}p_A{}^{\nu]}\delta f^A\,dS_{\mu\nu}\,, \tag{3.43}$$

with the odd form $dS_{\mu\nu}$ being defined by

$$dS_{\mu\nu} = (\partial_\mu \wedge \partial_\nu)\rfloor\, dx^1 \wedge \cdots \wedge dx^{n+1}\,. \tag{3.44}$$

It is instructive to present a direct derivation of equation (3.43) from Equation (3.41). For the purpose of illustration let f be a scalar field, with $\mathcal{D}_X f = \pounds_X f = X^\alpha f_{,\alpha}$. We have from (3.41):

$$\begin{aligned}\delta H &= \int_{\mathscr{S}} (\delta p^\mu X^\alpha f_{,\alpha} + p^\mu X^\alpha \delta f_{,\alpha} - X^\mu \delta L)\, dS_\mu \\ &= \int_{\mathscr{S}} \{\pounds_X f \delta p^\mu + \partial_\alpha[(p^\mu X^\alpha - X^\mu p^\alpha)\delta f] - \pounds_X p^\mu \delta f\}\, dS_\mu\,. \end{aligned} \tag{3.45}$$

Here we have used the field equations written in the form (3.21): $\delta L = \partial_\mu(p^\mu \delta f)$, and the fact that the Lie derivative $\pounds_X p^\mu$ of the vector density p^μ is given by the formula

$$\pounds_X p^\mu = \partial_\alpha(p^\mu X^\alpha) - X^\mu{}_{,\alpha} p^\alpha\,. \tag{3.46}$$

With $dS_{\mu\nu}$ defined as in equation (3.44) we have

$$\partial_\alpha\left[(p^\mu X^\alpha - X^\mu p^\alpha)\delta f\right] dS_\mu = d\left(p^{[\mu}X^{\alpha]}\delta f\, dS_{\mu\alpha}\right)\,. \tag{3.47}$$

If $\mathscr{S}$ is compact with differentiable boundary $\partial\mathscr{S}$, the last term in (3.45) can be integrated by parts, so that this equation can be rewritten as

$$-\delta H = \Omega_{\mathscr{S}}((\pounds_X p, \pounds_X f), (\delta p, \delta f)) + \int_{\partial\mathscr{S}} X^{[\mu}p^{\alpha]}\delta f\, dS_{\mu\alpha}\,, \tag{3.48}$$

with $\Omega_{\mathscr{S}}$ as in equation (3.37).

Formula (3.43) (or (3.48) in the particular case of a scalar field theory) can be used as a starting point of a definition of a Hamiltonian evolution of the field, provided the boundary integral is *made to vanish* by a suitable choice of boundary conditions. For example, when $\mathscr{S}$ is a compact hypersurface with boundary, a choice which is often possible consists of prescribing *a priori* the value of the Dirichlet data f^A on a timelike part of a boundary of a domain in space-time containing $\partial\mathscr{S}$. Performing a variation of f^A within this class of fields leads to δf^A that vanishes on $\partial\mathscr{S}$. In order to obtain a Hamiltonian dynamical system one needs then to make sure that those boundary conditions lead to a well defined dynamical system on an appropriately constructed, time-independent phase space. In some standard examples this procedure can be carried through. The main purpose of this paper is to describe various consistent ways of imposing such conditions on hypersurfaces $\mathscr{S}$ which extend up to $\mathscr{I}^+$, for a scalar field in Minkowski space-time and for the gravitational field.

Other boundary conditions — *e.g.*, Neumann conditions — are sometimes possible, but in general a modification of the Hamiltonian is needed. Suppose, for example, that $\mathscr{S}$ is compact with boundary, and that there is a well posed initial-boundary value problem with Neumann data, defined as the projection of $X^{[\mu}p_A{}^{\nu]}$ to the boundary $\partial\mathscr{S}$. One can then, as before, try to obtain Hamiltonian dynamics in a space, where the fields satisfy a Neumann condition at the timelike part of a boundary of a domain in space-time containing $\partial\mathscr{S}$. One has the trivial identity:

$$X^{[\mu}p_A{}^{\nu]}\delta f^A = \delta\left(X^{[\mu}p_A{}^{\nu]}f^A\right) - f^A X^{[\mu}\delta p_A{}^{\nu]} .$$

The Hamiltonian is then, eventually, obtained by transferring an integral involving $\delta\left(X^{[\mu}p_A{}^{\nu]}f^A\right)$ to the left hand side of (3.43).

$$\begin{aligned} -\delta\left(H + \int_{\partial\mathscr{S}} X^{[\mu}p_A{}^{\nu]}f^A\,dS_{\mu\nu}\right) = \\ \Omega_{\mathscr{S}}((\mathfrak{D}_X p_A{}^{\mu}, \mathfrak{D}_X f^A), (\delta p_A{}^{\mu}, \delta f^A)) - \int_{\partial\mathscr{S}} f^A X^{[\mu}\delta p_A{}^{\nu]}\,dS_{\mu\nu} . \end{aligned} \qquad (3.49)$$

Different boundary conditions correspond to different ways in which the portion of the field contained in $\mathscr{S}$ interacts with the rest of the world. The difference between different Hamiltonians obtained this way is similar to the difference between the *internal energy* and the *free energy* in thermodynamics: the first one describes the evolution of the system adiabatically insulated from the rest of the world whereas the second one describes the evolution of the same system put in the thermal bath. The two evolutions are different.

The Hamiltonian describing the evolution of a field configuration p in the direction of a vector field X defined over a hypersurface $\mathscr{S}$ may be obtained as an integral:

$$H(X,\mathscr{S},p) \equiv \int_{\mathscr{S}} H(X,p) . \qquad (3.50)$$

In the case where ψ arises from the flow on M of a vector field X, the integrand $H(X,p)$ is an odd n-form on M, defined by the formula

$$H(X,p) = H^{\mu}(X,p)\,\partial_{\mu}\rfloor\, dx^1 \wedge \cdots \wedge dx^{n+1} ,$$

with

$$H^{\mu}(X,p) \equiv p_A{}^{\mu}\mathfrak{D}_X f^A - X^{\mu}L . \qquad (3.51)$$

For a scalar field, or in the case when $\mathfrak{D}_X$ has been chosen to be a covariant derivative (along X) of a tensor field, the right hand side of (3.51) depends upon X in an algebraic (linear) way:

$$H^\mu(X,p) = \mathscr{T}^\mu{}_\nu(p) X^\nu , \tag{3.52}$$

where the proportionality coefficients $\mathscr{T}^\mu{}_\nu(p)$ form a tensor density, called usually *the energy-momentum tensor density*. In a more general case, when $\mathcal{D}_X$ is a Lie derivative of a tensor field, the right hand side of (3.51) depends also upon derivatives of the field X. In this case we have the following expansion:

$$H^\mu(X,p) = \mathscr{T}^\mu{}_\nu(p) X^\nu + \mathscr{T}^{\mu\lambda}{}_\nu(p) \nabla_\lambda X^\nu ,$$

which gives rise to the *second energy-momentum tensor density* $\mathscr{T}^{\mu\lambda}{}_\nu$, *etc.* When the (f^A)'s represent a connection field Γ in M, which is the case in the purely affine formulation of general relativity [91], the Lie derivative contains second derivatives of X, which gives rise to a "third energy-momentum tensor" in the above expansion.

3.4 Changes of Ψ and of the Lagrangian

Formula (3.51) is well adapted to the analysis of the dependence of the Hamiltonian upon the reference system Ψ. Indeed, different reference systems correspond to different operators $\mathcal{D}_X$. If $\mathcal{D}_X^{(1)}$ and $\mathcal{D}_X^{(2)}$ are two such operators, they differ by a vertical vector field on F:

$$\mathcal{D}_X^{(2)} f^A - \mathcal{D}_X^{(1)} f^A = v^A(f,x) .$$

It follows from (3.51), that the corresponding Hamiltonians differ by a function which is linear in momentum variables:

$$H^\mu_{(2)}(X,p) - H^\mu_{(1)}(X,p) = p_A{}^\mu v^A(f,x) .$$

To conclude this section, let us analyze what happens when we add to the Lagrangian $\mathbf{L}$ a complete divergence: this does not change the second order field equations for the variables f^A, but changes the definition of the momenta. To stay within the framework of first order variational principles, let the new Lagrangian be of the form:

$$\tilde{L}(j^1 f, x) := L(j^1 f, x) + \partial_\mu (R^\mu(f,x)) , \tag{3.53}$$

where R^μ is a vector-density-on-M-valued function on F. As a consequence, new momenta differ from the old ones by a non-homogeneous transformation:

$$\tilde{p}_A{}^\mu = \frac{\partial \tilde{L}}{\partial f^A{}_\mu} = p_A{}^\mu + \frac{\partial R^\mu}{\partial f^A} . \tag{3.54}$$

This is a canonical transformation and does not change the value of the form Ω_Σ given by formula (3.37). Nevertheless, such a transformation changes the Hamiltonian because it changes the connection in the bundle V^*F. Indeed, according to (3.14) we always have $\mathcal{D}_X p_A{}^\mu \equiv 0$ for a zero section $p_A{}^\mu \equiv 0$, whereas it is no longer true for the corresponding section $\tilde{p}_A{}^\mu = \frac{\partial R^\mu}{\partial f^A}$ after the transformation. The above phenomenon mimics the Galilei transformation in non-relativistic mechanics, where momenta undergo a non-homogeneous transformation which leads to a change in the Hamiltonian.

Assuming again that X is a vector field[9] on M, the new Hamiltonian can be calculated directly from formula (3.51):

$$\begin{aligned} \tilde{H}^\mu(X,\tilde{p}) &= \tilde{p}_A{}^\mu \mathcal{D}_X f^A - X^\mu \tilde{L} \\ &= H^\mu(X,p) + \frac{\partial R^\mu}{\partial f^A}\mathcal{D}_X f^A - X^\mu \partial_\nu (R^\nu(f,x)) \\ &= H^\mu(X,p) - \partial_\nu \left(X^\mu R^\nu - X^\nu R^\mu\right) - (\pounds_X R)^\mu \,, \end{aligned} \tag{3.55}$$

where we denote

$$(\pounds_X R)^\mu := X^\nu \frac{\partial}{\partial x^\nu}\left(R^\mu(f(x),x)\right) - \frac{\partial R^\mu}{\partial f^A}\mathcal{D}_X f^A - R^\nu \partial_\nu X^\mu - R^\mu \partial_\nu X^\nu \,. \tag{3.56}$$

This formula reduces to the standard expression for the Lie derivative of a vector density, when R^μ does not depend upon f^A. Otherwise, the second term on the right hand side of (3.56) modifies the derivative of R^μ from the first term, so that the result becomes a horizontal derivative in the bundle F.

The expression $\partial_\nu \left(X^\mu R^\nu - X^\nu R^\mu\right)$ leads to a boundary integral when integrated over a, say compact, hypersurface $\mathscr{S}$. If the fields satisfy Dirichlet boundary conditions, it gives rise to an additive constant in the Hamiltonian, which has no dynamical impact. In such a situation we see, therefore, that the difference between the two Hamiltonians is essentially given by a horizontal Lie derivative of R^μ in the direction of X^μ.

[9] Recall that formula (3.9) defines, for each τ, a section of TM over $\psi_\tau(\Sigma)$, but does not define a vector field over M in general. When X is not a vector field, formulae (3.55)–(3.56) do not make sense as such because of the partial derivatives of X in all directions appearing there, but remains correct when integrated upon $\psi_\tau(\Sigma)$: the only derivatives of X, which enter the integrated version of (3.55), are those which are tangent to $\psi_\tau(\Sigma)$.

4. Radiating scalar fields

4.1 Preliminaries

In this chapter we will consider the scalar wave equation on Minkowski space-time $\mathbb{R}^{3,1} \equiv (M,\eta) := (\mathbb{R}^4, \mathrm{diag}(-1,+1,+1,+1))$:

$$\Box_\eta f = 0\,. \tag{4.1}$$

Recall that scalar fields on M can be viewed as sections of a trivial bundle

$$\begin{array}{c} F = M \times \mathbb{R} \\ \Big\downarrow {\scriptstyle pr_{F\to M}} \\ M \end{array} \tag{4.2}$$

In this theory a natural choice for the Lagrangian is the one, which is manifestly invariant under Poincaré transformations:

$$L = -\frac{1}{2}\sqrt{-\det \eta_{\alpha\beta}}\ \eta^{\mu\nu}\partial_\mu f \partial_\nu f\ . \tag{4.3}$$

The canonical momentum p^μ is defined as in equation (3.18):

$$p^\mu = \frac{\partial L}{\partial f_{,\mu}} = -\sqrt{-\det \eta_{\alpha\beta}}\ \eta^{\mu\nu}\partial_\nu f\ . \tag{4.4}$$

Let $\mathscr{S}$ be any piecewise smooth hypersurface in M, let $p_1 = (p_1^\mu, f_1)$, $p_2 = (p_2^\mu, f_2)$ be two sections of the space-time phase bundle defined over $\mathscr{S}$; following Equation (3.37) we set

$$\Omega_{\mathscr{S}}(p_1,p_2) = \int_{\mathscr{S}} (p_1^\mu f_2 - p_2^\mu f_1)\, dS_\mu\,. \tag{4.5}$$

Here the p_a^μ's, $a = 1,2$, are calculated as in (4.4) with f replaced by f_a.

Our purpose here is to study the Hamiltonian aspects of such a theory in the radiation regime. It is convenient to use a conformal completion at $\mathscr{I}^+$ of M, constructed as follows: As a first step, let us represent M as a union $\cup_{\tau\in\mathbb{R}}\mathscr{S}_\tau$ of the following family of hyperbolae[1]:

[1] We choose hyperbolae for definiteness; the results of this section carry over immediately to any family of hypersurfaces which approach $\mathscr{I}^+$ in an appropriate way.

$$\mathscr{S}_\tau = \{x^0 = \tau + \sqrt{1 + (x^1)^2 + (x^2)^2 + (x^3)^2}\} .$$

Every $\mathscr{S}_\tau$ can be identified with an open unit ball $B(0,1) \approx S^2 \times [0,1)$ in $\mathbb{R}^3$:

$$x^0 = \tau + \sqrt{1 + r(\rho)^2} , \tag{4.6}$$

$$x^i = r(\rho) q^i , \qquad r = 2\rho/(1-\rho^2), \qquad \rho \in [0,1) , \tag{4.7}$$

$$q = (q^i) \in S^2 .$$

Following Penrose [105], we attach to M a boundary $\mathscr{I}^+$ by adding to the open ball $B(0,1)$ its boundary $\partial B(0,1) = S^2 \times \{\rho = 1\}$:

$$\widetilde{M} = \mathbb{R} \times \overline{B(0,1)} . \tag{4.8}$$

By an abuse of notation, we will represent $\widetilde{M}$ as $\mathbb{R} \times S^2 \times [0,1]$ using the spherical coordinates representation of the unit ball. The manifold with boundary $\widetilde{M}$ so defined is the standard conformal completion at $\mathscr{I}^+$ of M. Let $\overline{\mathscr{S}}_\tau$ be the closure of $\mathscr{S}_\tau$ in $\widetilde{M}$, set $S_\tau = \partial\overline{\mathscr{S}}_\tau = \overline{\mathscr{S}}_\tau \cap \mathscr{I}^+$, *cf.* Fig. 4.1.

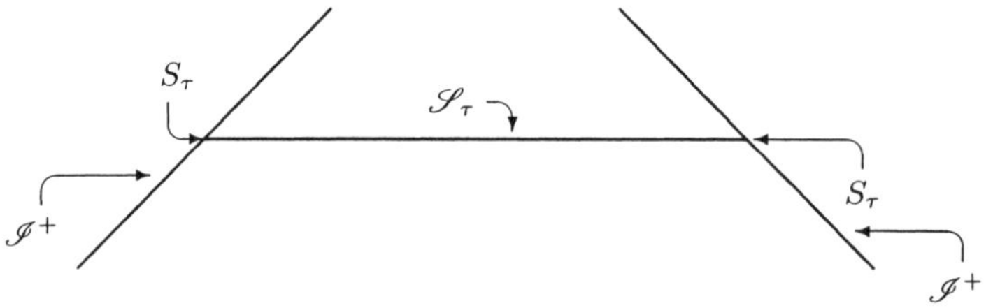

Fig. 4.1. A hyperboloid $\mathscr{S}_\tau$ in the conformally completed Minkowski space-time $\widetilde{M}$.

The S_τ's can be thought of as "spheres at infinity" on the hypersurfaces $\mathscr{S}_\tau$.

We wish to use the formalism of Chap. 3 to construct the Hamiltonian description of the field dynamics (4.1) associated with motions of hypersurfaces in M or in $\widetilde{M}$. While some technicalities related to the convergence of the integrals considered will arise, the results here are, essentially, a direct application of the formalism developed there. Scalar fields on $\widetilde{M}$ can be viewed as sections of a trivial bundle

$$\widetilde{F} = \widetilde{M} \times \mathbb{R} ,$$

the natural completion of (4.2). For any manifold Σ with embedding $i_\Sigma : \Sigma \to M$ (respectively $\widetilde{\Sigma}$ with embedding $i_{\widetilde{\Sigma}} : \widetilde{\Sigma} \to \widetilde{M}$) the pull-back $i_\Sigma!$ (respectively $i_{\widetilde{\Sigma}}!$) acts trivially on the scalar field:

$$\Phi_\Sigma = \Sigma \times \mathbb{R} , \; i_\Sigma! = i_\Sigma \times id_{\mathbb{R}} , \quad \text{respectively} \;\; \Phi_{\widetilde{\Sigma}} = \widetilde{\Sigma} \times \mathbb{R} , \; i_{\widetilde{\Sigma}}! = i_{\widetilde{\Sigma}} \times id_{\mathbb{R}} .$$

Here we use the notation Φ_Σ for the bundle Φ of equation (3.2), *etc.* Now in our applications the hypersurface $\widetilde{\Sigma}$ will consist of two pieces,

$$\widetilde{\Sigma} = \Sigma \cup \Sigma_{\mathscr{I}^+} , \tag{4.9}$$

with

$$i_\Sigma \equiv i_{\widetilde{\Sigma}}\big|_\Sigma , \quad i_\Sigma : \Sigma \to M , \qquad i_{\Sigma_{\mathscr{I}^+}} \equiv i_{\widetilde{\Sigma}}\big|_{\Sigma_{\mathscr{I}^+}} , \quad i_{\Sigma_{\mathscr{I}^+}} : \Sigma_{\mathscr{I}^+} \to \mathscr{I}^+ .$$

In all the cases considered below Σ will be an open ball of radius one, the image of which in M will be a hyperbola. Further, it will be natural to glue $\partial\Sigma$ to a connected component of $\partial\Sigma_{\mathscr{I}^+}$, so that $\widetilde{\Sigma}$ will actually be connected. Moreover, the motions ψ of Σ considered in this section will always arise from the flow of a vector field X defined on M; we mention that this will not be the case for the $\mathscr{I}^+$-part of the dynamics in the phase space considered in Sect. 4.5. It is natural to choose

$$\Psi = \psi \times id_{\mathbb{R}} . \tag{4.10}$$

If we set

$$\psi_{I\times\Sigma} \equiv \psi\Big|_{I\times\Sigma} , \qquad \psi_{I\times\Sigma_{\mathscr{I}^+}} = \psi\Big|_{I\times\Sigma_{\mathscr{I}^+}} , \tag{4.11}$$

with the obvious corresponding definitions for $\Psi_{I\times\Sigma}$ and $\Psi_{I\times\Sigma_{\mathscr{I}^+}}$, then

$$\Psi_{I\times\Sigma} = \psi_{I\times\Sigma} \times id_{\mathbb{R}} , \quad \Psi_{I\times\Sigma_{\mathscr{I}^+}} = \psi_{I\times\Sigma_{\mathscr{I}^+}} \times id_{\mathbb{R}} .$$

equation (3.10) gives then on M

$$\mathcal{D}_X f = X^\mu \partial_\mu f = \pounds_X f, \tag{4.12}$$

where $\pounds_X$ denotes a Lie derivative along X[2]. equation (3.41) leads thus to the following formula for the Hamiltonian:

$$H(X, \mathscr{S}, p) = \int_{\mathscr{S}} (p^\mu \pounds_X f - X^\mu L)\, dS_\mu . \tag{4.13}$$

Because we actually have

$$P_M = V^* F \otimes \tilde{\Lambda}^n M \approx F \oplus \tilde{\Lambda}^n M ,$$

where "$\approx$" denotes "isomorphic to", it is also natural to use the Lie derivative for the action of $\mathcal{D}_X$ on sections p of that part P_M of the space-time phase

[2] Actually, with the choice (4.10), the first equality in equation (4.12) will hold throughout $\widetilde{M}$ when X at $\psi(\tau, q)$ is defined by (3.9). This will, however, not be the Lie derivative as defined in standard textbooks on differential geometry, when X so constructed is not a vector field defined on $\widetilde{M}$. For scalar fields the distinction is only of esthetic nature; however, it becomes essential for general tensor fields.

bundle P which lies above M. With this choice the compatibility condition (3.42) holds.

It will be convenient to pass to the following coordinate system on $\widetilde{M}$, adapted to the construction above: $(y^\mu) = (\omega, y^A, \rho)$, with $y^0 := \omega := t - \sqrt{1+r^2} \in \mathbb{R}$ and $y^3 := \rho \in [0,1]$ being equal to the first and the last component in the decomposition $\widetilde{M} = \mathbb{R} \times S^2 \times [0,1]$, whereas $(y^A) = (\theta, \varphi)$, $A = 1,2$, are spherical coordinates on S^2. It induces the corresponding coordinate system (p^μ, f, y^μ) on P_M. We will also write $p^0 = p^\omega$. In terms of those coordinates the Minkowski metric reads

$$\eta = \eta_{\mu\nu}\, dx^\mu\, dx^\nu = \Omega^{-2}\tilde{g}_{\mu\nu}\, dy^\mu\, dy^\nu \;, \tag{4.14}$$

where

$$\tilde{g}_{\mu\nu}\, dy^\mu\, dy^\nu = -\left(\frac{1-\rho^2}{2}\right)^2 d\omega^2 - 2\rho\, d\omega\, d\rho + d\rho^2 + \rho^2 \breve{h}_{AB}\, dy^A\, dy^B \;, \tag{4.15}$$

$$\Omega = \frac{1-\rho^2}{2} \;, \tag{4.16}$$

and $\breve{h}_{AB}\, dy^A\, dy^B = d\theta^2 + \sin^2\theta\, d\varphi^2$. For future reference we note the formulae

$$\sqrt{|\det \eta|} = \frac{8\rho^2(1+\rho^2)}{(1-\rho^2)^4}\sin\theta \;, \tag{4.17}$$

$$\begin{aligned}\eta^{\mu\nu}\partial_\mu\partial_\nu &= \left(\frac{1-\rho^2}{1+\rho^2}\right)^2 \left[-\partial_\omega^2 - 2\rho\partial_\omega\partial_\rho + \left(\frac{1-\rho^2}{2}\right)^2 \partial_\rho^2\right] \\ &\quad + \left(\frac{1-\rho^2}{2\rho}\right)^2 \breve{h}^{AB}\partial_A\partial_B \;, \end{aligned}\tag{4.18}$$

$$\breve{h}^{AB}\partial_A\partial_B = \partial_\theta^2 + \sin^{-2}\theta\partial_\varphi^2 \;. \tag{4.19}$$

Now, we are ready to impose boundary conditions on $\mathscr{I}^+$, or subsets thereof, for solutions of (4.1). Let $I \subset \mathbb{R}$, we define

$$\mathscr{I}^+_I = \cup_{u\in I} S_u \;; \tag{4.20}$$

thus $\mathscr{I}^+_{\mathbb{R}} = \mathscr{I}^+$, and, *e.g.*, $\mathscr{I}^+_{(\tau_-,\tau)} = \cup_{u\in(\tau_-,\tau)} S_u$ is a smooth null hypersurface contained in $\mathscr{I}^+$ with boundary $S_\tau \cup S_{\tau_-}$. Throughout this paper we will be interested only in those solutions of (4.1), which extend smoothly to $\mathscr{I}^+_{(\tau_-,\tau_1]}$ in the following sense: Under the conformal transformation of the metric,

$$\eta_{\mu\nu} \longrightarrow \tilde{g}_{\mu\nu} = \Omega^2 \eta_{\mu\nu} \;, \tag{4.21}$$

Equation (4.1) becomes

$$\Box_{\tilde{g}}\tilde{f} + \frac{1}{6}R(\tilde{g})\tilde{f} = 0\,, \tag{4.22}$$

with

$$\tilde{f} = \frac{f}{\Omega}\,,$$

where $R(\tilde{g})$ is the Ricci scalar of the metric $\tilde{g}$. We shall require that the function $\tilde{f}$ be extendable by continuity to a smooth function on $\widetilde{M}_{(\tau_-,\tau]} = \cup_{\tau\in(\tau_-,\tau]}\overline{\mathscr{S}}_\tau$. It can be seen, *e.g.* using Taylor expansions around $\mathscr{I}^+$, that this is equivalent to smooth extendability of f to $\mathscr{I}^+_{(\tau_-,\tau_1]}$ together with the vanishing on $\mathscr{I}^+_{(\tau_-,\tau_1]}$ of the extension so obtained. By an abuse of notation we shall use the same symbol $\tilde{f}$ to denote this extension. See the Appendix B for a discussion of how to construct such solutions.

4.2 Energy: convergence of integrals

Suppose, thus, that f, f_a and δf are smooth on $\widetilde{M}$ with $f|_{\mathscr{I}^+} = f_a|_{\mathscr{I}^+} = \delta f|_{\mathscr{I}^+} = 0$. In the coordinate system (y^μ) used above we can write

$$f = \Omega\tilde{f}\,, \quad f_a = \Omega\tilde{f}_a\,, \quad \delta f = \Omega\delta\tilde{f}\,, \tag{4.23}$$

for some functions $\tilde{f}$, $\tilde{f}_a$, $\delta\tilde{f}$, smooth on $\widetilde{M}$. For $\varepsilon > 0$ let $\mathscr{S} = \mathscr{S}_{\tau,\varepsilon} \equiv \mathscr{S}_\tau \cap \{\rho \le 1-\varepsilon\}$. We would like to pass with ε to zero in (4.5) and (4.13), as well as in the variational equation (3.48). This requires defining the phase spaces under consideration, and then checking the convergence of the integrals involved. We will actually proceed the other way round, first checking the convergence and/or the vanishing of the relevant integrals, to obtain a hint how the phase spaces should be defined. This will eventually lead to a definition of three possible phase spaces in Sects. 4.3–4.5 below.

In the coordinates (ω, y^A, ρ) the only p^μ terms, which contribute to (4.5) and (4.13), are those which involve p^ω. From (4.17)–(4.18) and (4.4) we have

$$\Omega p^\omega = \frac{2\rho^2\sin\theta}{(1+\rho^2)}\left[\frac{\partial\tilde{f}}{\partial\omega} + \rho\frac{\partial\tilde{f}}{\partial\rho} - \frac{2\rho^2}{1-\rho^2}\tilde{f}\right]. \tag{4.24}$$

Now in (4.5) we have $f_a = \frac{1-\rho^2}{2}\tilde{f}_a = O(1-\rho)$ while $\frac{1-\rho^2}{2}p^\omega$ blows up as $1/(1-\rho)$ at $\mathscr{I}^+$, and a naive inspection of (4.5) suggests that the integral over $\mathscr{S}_{\tau,\varepsilon}$ will diverge in the limit. However, because of the antisymmetry of $\Omega_{\mathscr{S}}$ in f_a, the offending term $2\rho^2\tilde{f}/(1-\rho^2)$ from (4.24) drops out when inserted in (4.5). It is convenient to define the scalar density $\tilde{p} \equiv \tilde{p}^\omega$ on $\mathscr{S}_\tau$, which in the coordinate system (ρ, y^A) equals

$$\tilde{p} = \frac{2\rho^2\sin\theta}{(1+\rho^2)}\left[\frac{\partial\tilde{f}}{\partial\omega} + \rho\frac{\partial\tilde{f}}{\partial\rho}\right]. \tag{4.25}$$

With this definition one obtains

$$\Omega_{\mathscr{S}_\tau}\left((\delta_1\tilde{p},\delta_1\tilde{f}),(\delta_2\tilde{p},\delta_2\tilde{f})\right) \equiv \int_{\mathscr{S}_\tau}(\delta_1\tilde{p}\,\delta_2\tilde{f}-\delta_2\tilde{p}\,\delta_1\tilde{f})\,d\rho\,d\theta\,d\varphi\,, \qquad (4.26)$$

with $(\tilde{p}_a,\tilde{f}_a)$, $a=1,2$, related to $\tilde{f}_a$ by the formulae above. (4.26) is obviously well defined when $\delta_1\tilde{f}$ and $\delta_2\tilde{f}$ are smooth on $\widetilde{M}$.

We see that the parameterization of the phase space in terms of the quantities $(\tilde{p},\tilde{f})$, instead of the original quantities (p,f) enables us to rewrite all the integrals over hyperboloids in terms of manifestly finite quantities. Formula (4.26) shows that this reparameterization is, loosely speaking, a canonical transformation.

We note that the quantity $\tilde{p}$ could have been derived as a "momentum canonically conjugate to $\tilde{f}$" from the following, modified Lagrangian:

$$\begin{aligned}\tilde{L} &:= L+\frac{1}{2}\partial_\nu\left(\sqrt{-\eta}\eta^{\mu\nu}\Omega_{,\mu}\Omega^{-1}f^2\right)\\ &= -\frac{1}{2}\sqrt{-\tilde{g}}\tilde{g}^{\mu\nu}\tilde{f}_\mu\tilde{f}_\nu+\frac{1}{12}\sqrt{-\tilde{g}}R(\tilde{g})\tilde{f}^2\,. \end{aligned} \qquad (4.27)$$

Indeed, putting

$$\begin{aligned}\tilde{p}^\mu &:= \frac{\partial\tilde{L}}{\partial\tilde{f}_{,\mu}} = -\sqrt{-\tilde{g}}\tilde{g}^{\mu\nu}\partial_\nu\tilde{f}\\ &= -\Omega^2\sqrt{-\eta}\eta^{\mu\nu}\left[\frac{1}{\Omega}\partial_\nu f-\frac{f}{\Omega^2}\partial_\nu\Omega\right]\\ &= \Omega p^\mu+\frac{1}{2}\tilde{f}\sqrt{-\eta}\eta^{\mu\nu}\partial_\nu\Omega^2\,, \end{aligned} \qquad (4.28)$$

we obtain $\tilde{p}=\tilde{p}^\omega$ and it is precisely the last, non-homogeneous term which was dropped out when passing from (4.24) to (4.25). For fields $\tilde{f}$ considered here, the function $\tilde{L}$ may be extended to $\mathscr{I}^+$ by continuity, which is not the case for L. We stress, however, that we *do not* change the Lagrangian, which would imply change of the reference system for momenta and — consequently — change of the Hamiltonian (see formula (3.54) and the discussion below). In particular, the derivative $\mathfrak{D}_X$ is always calculated with respect to the previous structure, carried by the variable f:

$$\mathfrak{D}_X\tilde{f} \equiv \frac{1}{\Omega}\mathfrak{D}_X f = \frac{1}{\Omega}\pounds_X(\Omega\tilde{f}) = \pounds_X\tilde{f}+\tilde{f}\frac{1}{\Omega}\pounds_X\Omega\,.$$

(The leftermost equality above is the transformation law for a connection.)

Let us pass to the analysis of the convergence of the integral (4.13). Consider, first, the case when X is the vector field generating time translations:

$$X=\frac{\partial}{\partial x^0}=\frac{\partial}{\partial y^0}=\frac{\partial}{\partial\omega}\,. \qquad (4.29)$$

The corresponding Hamiltonian is the field energy. With X as in equation (4.29) the multiplication by Ω commutes with $\pounds_X$ because $\pounds_X \Omega \equiv 0$, which simplifies somewhat the analysis. In this case, the "space density"

$$p^\omega \pounds_X f = \frac{1-\rho^2}{2} p^\omega \pounds_X \tilde{f}$$

blows up as $1/(1-\rho)$ at $\mathscr{I}^+$. However, there is a cancellation with a similar term in L which leads to

$$\begin{aligned} H^\omega(X,p) &= p^\omega \pounds_X f - L \\ &= \frac{\rho^2 \sin\theta}{(1+\rho^2)} \left\{ \left(\frac{\partial \tilde{f}}{\partial \omega}\right)^2 + \left[\partial_\rho \left(\frac{1-\rho^2}{2}\tilde{f}\right)\right]^2 \right\} \\ &\quad + \frac{1+\rho^2}{4} \sin\theta \breve{h}^{AB} \partial_A \tilde{f} \partial_B \tilde{f} \,. \end{aligned} \tag{4.30}$$

It follows that we can safely pass to the limit $\varepsilon \to 0$ in $H(X, \mathscr{S}_{T,\epsilon}, p)$ to obtain a finite and well defined $H(X, \mathscr{S}_T, p)$ when $\tilde{f} \in C^\infty(\tilde{M})$. Moreover, as we have already seen in the derivation of formula (4.26), the non-homogeneous terms coming from $\delta\tilde{p}$ and from $\pounds_X \tilde{p}$ cancel when inserted into the form $\Omega_{\mathscr{S}}$:

$$\pounds_X f \delta p - \pounds_X p \delta f = \pounds_X \tilde{f} \delta \tilde{p} - \pounds_X \tilde{p} \delta \tilde{f} \,,$$

(as $X = \partial/\partial\omega$, we actually have $\pounds_X \tilde{p} = \tilde{p}_{,\omega}$). This implies that the potentially dangerous terms in (3.45) cancel, so that (3.48) can be rewritten as

$$\begin{aligned} -\delta H(X, \mathscr{S}_T, p) &= \int_{\mathscr{S}_T} (\pounds_X \tilde{p}\, \delta \tilde{f} - \pounds_X \tilde{f}\, \delta\tilde{p}) \\ &\quad + \lim_{\varepsilon\to 0} \int_{\partial \mathscr{S}_{T,\varepsilon}} \Omega X^{[\mu} p^{\alpha]} \delta \tilde{f}\, dS_{\mu\alpha} \,, \end{aligned} \tag{4.31}$$

with

$$H(X, \mathscr{S}_T, p) := \int_0^1 d\rho \int_{S^2} H^\omega(X,p)\, d\theta\, d\varphi \,. \tag{4.32}$$

Also the boundary integral over $\partial\mathscr{S}_{T,\varepsilon}$ in (4.31) converges when $\varepsilon \to 0$ because the only term p^μ, which gives a non-vanishing contribution, is the one involving p^ρ, which by (4.15)-(4.19) reads

$$\Omega p^\rho = \frac{2\rho^2 \sin\theta}{1+\rho^2} \left\{ \rho \frac{\partial \tilde{f}}{\partial \omega} - \frac{1-\rho^2}{2} \partial_\rho \left[\frac{1-\rho^2}{2} \tilde{f} \right] \right\} . \tag{4.33}$$

On the boundary $\partial\mathscr{S} = \{\rho = 1\}$ it reduces to

$$\tilde{p}^\rho = \Omega p^\rho = \sin\theta \frac{\partial \tilde{f}}{\partial \omega} \,. \tag{4.34}$$

Hence, equations (4.31)–(4.33) yield in the limit:

$$-\delta H(X,\mathscr{S}_\tau,p) = \Omega_{\mathscr{S}_\tau}((\pounds_X\tilde{p},\pounds_X\tilde{f}),(\delta\tilde{p},\delta\tilde{f})) + \int_{\partial\mathscr{S}_\tau}\frac{\partial\tilde{f}}{\partial\omega}\delta\tilde{f}\sin\theta\,d\theta\,d\varphi \quad (4.35)$$

A similar analysis may be performed for that part of $\widetilde{\Sigma}$, which is mapped by ψ to the boundary $\mathscr{I}^+$ of $\widetilde{M}$. The relevant field data will be described by the value on $\mathscr{I}^+$ of (the extended field) $\tilde{f}$, which we denote by χ:

$$\chi(\omega,y^A) \equiv \tilde{f}(\omega,\rho=1,y^A), \quad \delta\chi(\omega,y^A)\equiv\delta\tilde{f}(\omega,\rho=1,y^A). \quad (4.36)$$

Formula (4.34) for the corresponding momentum $\tilde{p}^\rho$ is a (Hamiltonian) constraint in the phase space of the field data on $\mathscr{I}^+$. Equation (4.5) yields the following formula for the symplectic structure in the space of functions χ:

$$\begin{aligned}\Omega_{\mathscr{I}^+_I}(\delta_1\chi,\delta_2\chi) &\equiv \lim_{\epsilon\to0}\int_{\{\rho=1-\epsilon\}\times I\times S^2}(\delta_1 p^\mu\delta_2 f-\delta_2 p^\mu\delta_1 f)\,dS_\mu\\ &= \lim_{\epsilon\to0}\int_{\tau\in I}d\tau\int_{S^2}(\delta_1 p^\rho\delta_2 f-\delta_2 p^\rho\delta_1 f)\Big|_{\rho=1-\epsilon}\sin\theta\,d\theta\,d\varphi\\ &= \int_{\tau\in I}d\tau\int_{S_\tau}\left(\frac{\partial\delta_1\chi}{\partial\omega}\delta_2\chi-\frac{\partial\delta_2\chi}{\partial\omega}\delta_1\chi\right)\sin\theta\,d\theta\,d\varphi\,. \quad (4.37)\end{aligned}$$

Here we have used equation (4.33). The corresponding Hamiltonian density reads:

$$H^\rho(X,\chi) = (p^\rho\pounds_X f - X^\rho L)\Big|_{\mathscr{I}^+} = (\tilde{p}^\rho\pounds_X\tilde{f})\Big|_{\mathscr{I}^+} = \sin\theta\left(\frac{\partial\chi}{\partial\omega}\right)^2. \quad (4.38)$$

In this context equation (3.48) takes the following form

$$\begin{aligned}-\delta H(X,\mathscr{I}^+_{[\tau_-,\tau]},\chi) &= \Omega_{\mathscr{I}^+_{[\tau_-,\tau]}}(\pounds_X\chi,\delta\chi)\\ &\quad+\int_{S_\tau}\frac{\partial\chi}{\partial\omega}\delta\chi\sin\theta\,d\theta\,d\varphi-\int_{S_{\tau_-}}\frac{\partial\chi}{\partial\omega}\delta\chi\sin\theta\,d\theta\,d\varphi\,, \quad (4.39)\end{aligned}$$

where

$$H(X,\mathscr{I}^+_{[\tau_-,\tau]},\chi) = \int_{\tau_-}^{\tau}d\omega\int_{S^2}H^\rho(X,\chi)\,d\theta\,d\varphi\,.$$

Formula (4.39) has been derived in a rather general formalism, in which the presence of constraints — or lack thereof — was hidden in the formalism. It is instructive to rederive it directly for the system constrained by equation (4.34); we have

$$\begin{aligned}\delta\left(\frac{\partial\chi}{\partial\omega}\right)^2 &= 2\frac{\partial\chi}{\partial\omega}\frac{\partial\delta\chi}{\partial\omega} = \frac{\partial\chi}{\partial\omega}\frac{\partial\delta\chi}{\partial\omega}+\frac{\partial}{\partial\omega}\left(\frac{\partial\chi}{\partial\omega}\delta\chi\right)-\frac{\partial^2\chi}{\partial\omega^2}\delta\chi\\ &= \frac{\partial\delta\chi}{\partial\omega}\pounds_X\chi-\frac{\partial\pounds_X\chi}{\partial\omega}\delta\chi+\frac{\partial}{\partial\omega}\left(\frac{\partial\chi}{\partial\omega}\delta\chi\right), \quad (4.40)\end{aligned}$$

which gives precisely (4.39) after integration.

Gluing together $\partial\mathscr{S}_\tau$ with the corresponding piece of $\partial\mathscr{I}^+_{[\tau_-,\tau]}$, we finally obtain:

$$-\delta H(X, \mathscr{S}_\tau \cup \mathscr{I}^+_{[\tau_-,\tau]}, p) = \Omega_{\mathscr{S}_\tau}\Big((\pounds_X\tilde{p}, \pounds_X\tilde{f}), (\delta\tilde{p}, \delta\tilde{f})\Big) + \Omega_{\mathscr{I}^+_{[\tau_-,\tau]}}(\pounds_X\chi, \delta\chi) + \int_{S_{\tau_-}} \frac{\partial\chi}{\partial\omega}\delta\chi \sin\theta \, d\theta \, d\varphi \,. \tag{4.41}$$

4.3 The phase space $\mathscr{P}_{(-\infty,0]}$

Throughout the discussion above we have assumed that solutions of the wave equation, which are smooth on $\widetilde{M}$, exist, this question is discussed in detail in Appendix B. Using the information contained there, we are ready now to pass to the description of various phase spaces in which hyperboloids moving in $\widetilde{M}$ induce a Hamiltonian dynamical system. The first such phase space we consider is that of appropriate data (Cauchy on $\mathscr{S}$, and characteristic on $\mathscr{I}^+$) for solutions of the wave equation on $\mathbb{R}^{1,3}$ such that $\tilde{f}|_{\mathscr{I}^+}$ is smooth and ω-independent for large negative values of ω. We shall follow the philosophy adopted in Chap. 3, and use different symbols for fields defined on $\widetilde{M}$, as compared to those on the model space $\Sigma \times \mathbb{R}$. More precisely, let $\mathscr{P}_{(-\infty,0]}$ denote the space of triples $(\tilde{\pi}, \tilde{\phi}, c)$ such that[3]:

- $\tilde{\pi}$ is a smooth-up-to-boundary function on $\overline{B(0,1)}$ (recall that we are identifying $B(1)$ with a unit hyperboloid in $\mathbb{R}^{1,3}$ using equations (4.6)–(4.7));
- $\tilde{\pi}$ is a density on $B(0,1)$, which is proportional to the two-volume density on S^2 (in spherical coordinates (θ,φ) equal to $\sin\theta$), with the proportionality coefficient smooth-up-to-boundary on $\overline{B(0,1)}$;
- c is a smooth function on $(-\infty,0] \times S^2$ (which should be thought of here as being a subset of $\mathscr{I}^+$), with $\partial c/\partial\omega$ — compactly supported;
- $(\tilde{\pi}, \tilde{\phi}, c)$ satisfy the corner conditions to all orders (*cf.* Appendix B). In particular, the continuity condition reads:

  $$\tilde{\phi}\,|_{\rho=1} = c|_{\omega=0} \,, \tag{4.42}$$

  and the differentiability condition implied by (4.25) reads:

  $$\left(\frac{\tilde{\pi}}{\sin\theta} - \frac{\partial\tilde{\phi}}{\partial\rho}\right)\Bigg|_{\rho=1} = \frac{\partial c}{\partial\omega}\bigg|_{\omega=0} \tag{4.43}$$

  (a finite number of corner conditions would suffice if finite differentiability of $\Omega^{-1}f$ on $\widetilde{M}$ is only required).

[3] A somewhat similar phase space has already been considered in [9] and [83].

Loosely speaking, the couple $(\tilde{\pi}, \tilde{\phi})$ can be thought of as the value of the couple $(\tilde{p}, \tilde{\phi})$ on the hyperboloid $\mathscr{S}_0$, while c can be thought of as the value of $\tilde{\phi}$ on $\mathscr{I}^+_{(-\infty,0]}$; *cf.* equation (4.49) below.

Given $(\tilde{\pi}, \tilde{\phi}, c)$, one solves the wave equation (4.22) with initial data

$$(f\Omega^{-1})|_{\mathscr{S}_0} = \tilde{\phi}\,, \tag{4.44}$$

$$(\nabla^\omega f)|_{\mathscr{S}_0} = \frac{1}{\sqrt{-\det \eta}} p^\omega(\tilde{\pi}, \tilde{\phi})\,, \tag{4.45}$$

$$(\lim_{\Omega\to 0} f\Omega^{-1})\Big|_{\mathscr{I}^+_{(-\infty,0]}} = c\,, \tag{4.46}$$

with $p^\omega(\tilde{\pi}, \tilde{\phi})$ — a solution of equation (4.28) with $\mu = \omega$, $\tilde{f} = \tilde{\phi}$ and $\tilde{p}^\omega = \tilde{\pi}$. Then $(\tilde{\pi}_t, \tilde{\phi}_t)$ will be calculated from f using equations (4.44)–(4.45) with $\mathscr{S}_0$ replaced by $\mathscr{S}_t$, while c_t is obtained from

$$c_t(\omega, \theta, \varphi) = \left(\lim_{\Omega\to 0} (\tilde{f}\Omega^{-1})\right)(\omega - t, \theta, \varphi)\,. \tag{4.47}$$

We then set

$$T_t((\tilde{\pi}, \tilde{\phi}, c)) \equiv (\tilde{\pi}_t, \tilde{\phi}_t, c_t)\,. \tag{4.48}$$

In summary, in local coordinates (ω, x^A, ρ) on $\widetilde{M}$ we have

$$\begin{aligned} \tilde{\pi}_t(x^A, \rho) &= \tilde{p}(t, x^A, \rho)\,, \\ \tilde{\phi}_t(x^A, \rho) &= \tilde{f}(t, x^A, \rho)\,, \\ c_t(\omega, x^A) &= \chi(\omega - t, x^A)\,. \end{aligned} \tag{4.49}$$

One expects that the condition, that $\partial c/\partial \omega$ is compactly supported, implies that at fixed Minkowskian time and for large r the field f behaves as $c(\theta, \varphi)/r + O(r^{-2})$, with $\partial f/\partial t = O(r^{-2})$, but no rigourous estimates of this kind exist so far.

Note that in (4.41) and (4.37) the interval $[\tau_-, \tau]$ can be replaced by $(-\infty, 0]$ since $\frac{\partial c}{\partial \omega}$ and $\frac{\partial \delta c}{\partial \omega}$ have compact support. Further, the integral over S_{τ_-} in (4.41) vanishes for large negative τ_-, so that it will give no contribution when passing to the limit $\tau_- \to -\infty$. It follows from equation (4.41) with $[\tau_-, \tau]$ replaced by $(-\infty, 0]$ that the dynamics is Hamiltonian in the sense of Sect. 2.1, which we desired to show. The Hamiltonian $H(\cdot)$ as defined by equation (2.2) equals

$$\begin{aligned} H(\frac{\partial}{\partial\omega}, \mathscr{S}_0 \cup \mathscr{I}^+_{(-\infty,0]}, \cdot) &= H(\frac{\partial}{\partial\omega}, \mathscr{S}_0, \cdot) + H(\frac{\partial}{\partial\omega}, \mathscr{I}^+_{(-\infty,0]}, \cdot) \\ &= \int_0^1 d\rho \int_{S^2} H^\omega(X, p)\, d\theta\, d\varphi \\ &\quad + \int_{-\infty}^0 d\omega \int_{S^2} \sin\theta \left(\frac{\partial c}{\partial \omega}\right)^2 d\theta\, d\varphi\,, \end{aligned} \tag{4.50}$$

with $H^\omega(X,p)$ given by equation (4.30). This quantity is nothing but the integral over $\mathscr{S}_0 \cup \mathscr{I}^+_{(-\infty,0]}$ of the standard energy-momentum tensor $\mathscr{T}^\mu{}_\nu$ of a massless scalar field (*cf.* equation (3.52)) contracted with the Killing vector $X = \frac{\partial}{\partial\omega}$. Since $\mathscr{T}^\mu{}_\nu X^\nu$ has vanishing divergence, one expects that the numerical value of H equals the "ADM energy" H_{ADM} of f (that is, H defined by equation (2.3), with $\mathbb{R}^3$ there being thought of as a hypersurface of constant standard Minkowskian time). To establish such an equality one would, however, need to have better control of f near i^0 than what follows from continuity of $\tilde{f}$ on $\widetilde{M}$. We simply note that for fields in $\mathscr{P}_{(-\infty,0]}$ we have

$$H_{ADM} \leq H$$

(in particular H_{ADM} will be finite), and that equality holds if *e.g.* $f|_{t=0}$ and $\frac{\partial f}{\partial t}|_{t=0}$ are compactly supported. Since $\mathscr{P}_{(-\infty,0]}$ is a vector space, it is path connected, which implies that H is uniquely defined up to a constant. The freedom of choice of that constant can be gotten rid of by requiring that H vanishes on the trivial field configuration $(\tilde{\pi},\tilde{\phi},c) \equiv (0,0,0)$.

4.4 The phase space $\mathscr{P}_{[-1,0]}$

In some situations it might be convenient to consider solutions of the field equations, which are defined only to the future of a prescribed hyperboloid $\mathscr{S}_{\tau_0}$, or on a neighbourhood thereof. This will be particularly apparent in the case of the gravitational field, where no global existence theorem of the kind we have for the linear massless scalar field holds. Our next phase space $\mathscr{P}_{[-1,0]}$ corresponds to such a situation: $\mathscr{P}_{[-1,0]}$ is defined as the space of triples $(\tilde{\pi},\tilde{\phi},c)$, where $(\tilde{\pi},\tilde{\phi})$ are as in $\mathscr{P}_{(-\infty,0]}$, while the c's are smooth functions on $[-1,0]\times S^2$ such that $\partial c/\partial\omega = 0$ in a neighbourhood of $\{-1\}\times S^2$. We can extend c to a smooth function $\hat{c}$ defined on $(-\infty,0]$, and then set $T_t(\tilde{\pi},\tilde{\phi},c) = (\tilde{\pi}_t,\tilde{\phi}_t,\hat{c}_t|_{[-1,0]})$, where $(\tilde{\pi}_t,\tilde{\phi}_t,\hat{c}_t)$ have been defined as in $\mathscr{P}_{(-\infty,0]}$ using $\hat{c}$. Suppose, then, that $(\tilde{\pi},\tilde{\phi},c) \in \mathscr{P}_{[-1,0]}$ is such that $\partial c/\partial\omega = 0$ on $[-1,\alpha]\times S^2 \subset [-1,0]\times S^2$, and that α is the largest number in $[0,1]$ with this property. We note that the $\mathscr{I}^+$ component of the dynamics on $\mathscr{P}$, that is, the dynamics of $\hat{c}$, is essentially a shift in the first variable of $\hat{c}$, so that we will have $\partial\hat{c}_t(\omega,y^A)/\partial\omega = 0$ on $[-1,-1+\alpha-t]\times S^2$ as long as $t\in[0,\alpha]$. For $t>\alpha$ the function $\hat{c}_t$ will in general not be constant in ω on $(-\infty,-1]$, and will therefore not correspond to a function in $\mathscr{P}_{[-1,0]}$ any more. Thus T_t is only defined for an interval of positive times depending upon the point $(\tilde{\pi},\tilde{\phi},c)$ of $\mathscr{P}_{[-1,0]}$, hence is only a local dynamical system. While this might be considered as an unsatisfactory feature, we emphasize that this is typical for those nonlinear theories (such as Einstein equations) where blow up in finite time occurs, so that this kind of behaviour has to be accepted. Formula (4.41) with $[\tau_-,\tau] = [-1,0]$ shows that the dynamics is Hamiltonian, with

$$H(\cdot) = H(\frac{\partial}{\partial\omega}, \mathscr{S}_0 \cup \mathscr{I}^+_{[-1,0]}, \cdot)\ , \tag{4.51}$$

and

$$\Omega\left((\delta_1\tilde{\pi}, \delta_1\tilde{\phi}, \delta_1 c), (\delta_2\tilde{\pi}, \delta_2\tilde{\phi}, \delta_2 c)\right)$$
$$\equiv \Omega_{\mathscr{S}_0}\left((\delta_1\tilde{p}, \delta_1\tilde{f}), (\delta_2\tilde{p}, \delta_2\tilde{f})\right) + \Omega_{\mathscr{I}^+_{[-1,0]}}(\delta_1\chi, \delta_2\chi)\ . \tag{4.52}$$

Here the $(\tilde{p}_a, \tilde{f}_a, \chi_a)$'s, $a = 1, 2$, have been identified with the $(\tilde{\pi}_a, \tilde{\phi}_a, c_a)$'s using Equation (4.49) with $t = 0$. This follows from the fact that $\frac{\partial c}{\partial \omega}\big|_{S_{-1}} = 0$, so that the integral over S_{τ_-} in (4.41) vanishes. Since $\mathscr{T}^\mu{}_\nu X^\nu$ is divergence-free we have

$$H(\cdot) = H(\frac{\partial}{\partial\omega}, \mathscr{S}_{-1}, \cdot)\ , \tag{4.53}$$

a quantity which we call the Trautman–Bondi energy of f on the hyperboloid $\mathscr{S}_{-1}$. We note, however, that (4.52) is explicitly expressible in terms of $(\tilde{\pi}, \tilde{\phi}, c)$, while (4.53) is not: the calculation of (4.53) requires solving backwards in time the wave equation with the data $(\tilde{\pi}, \tilde{\phi}, c)$ to obtain the relevant fields on $\mathscr{S}_{-1}$. $\mathscr{P}_{[-1,0]}$ is again a vector space, so that the Hamiltonian H is defined up to an additive constant. There is, however, an *essential ambiguity* which arises here if we try to represent H as a functional defined over all field configurations, not only those that have vanishing time derivatives in a neighbourhood of S_{-1}. Indeed, we can add to H any functional of c and a finite number of its derivatives at S_{-1}, which has the property that it vanishes when $\frac{\partial c}{\partial \omega}\big|_{S_{-1}} = 0$. From the phase space point of view this will be identical with H, but will not coincide with it for general fields which are not in $\mathscr{P}_{[-1,0]}$. This is discussed in more detail in Sect. 4.6.

4.5 The phase space $\widehat{\mathscr{P}}_{[-1,0]}$

To obtain the last Hamiltonian description of the dynamics, we have "killed" the boundary term in (4.41) by imposing the condition $\partial c/\partial\omega|_{S_{-1}} = 0$ in a neighbourhood of S_{-1}. Let us show now that a Hamiltonian description of the field dynamics on $\mathscr{I}^+$ is also possible without imposing any such conditions, provided one accepts to use *time-dependent* Hamiltonians. In the notation of equation (4.9), let

$$\Sigma_{\mathscr{I}^+} = [-1, 0] \times S^2\ ,$$

and let

$$\psi_{I\times\Sigma_{\mathscr{I}^+}} : (-1, \infty) \times \Sigma_{\mathscr{I}^+} \to \mathscr{I}^+$$

be any family of maps, which act trivially on the S^2 factor of $\Sigma_{\mathscr{I}^+}$,

$$\psi_{I\times\Sigma_{\mathscr{I}^+}}(\tau, \omega, x^A) = (\sigma_\tau(\omega), x^A) \in \mathbb{R} \times S^2 \approx \mathscr{I}^+\ .$$

Here $(\sigma_t)_{t>-1}$ is any smooth family of smooth orientation preserving diffeomorphisms from $[-1,0]$ to $[-1,t]$, with $\sigma_0(u)=u$. Let $\widehat{\mathscr{P}}_{[-1,0]}$ be the space of triples $(\tilde{\pi},\tilde{\phi},c)$ defined as for $\mathscr{P}_{[-1,0]}$, except that we do not impose the condition of $\partial c/\partial\omega$ at S_{-1}. Given a "time" $t>-1$ and a point $(\tilde{\pi},\tilde{\phi},c)\in\widehat{\mathscr{P}}_{[-1,0]}$ we construct for any $s>-1$ the evolution map $T_{s,t}$ (depending upon σ) as follows: Let f be the solution of the massless wave equation defined on the causal future $J^+(\overline{\mathscr{S}}_{-1},\widetilde{M})$ of $\overline{\mathscr{S}}_{-1}$ in $\widetilde{M}$ such that $\tilde{\phi}$ and $\tilde{\pi}$ correspond to the initial data for f on $\mathscr{S}_t$ in the sense already described, *cf.* equations (4.23) and (4.25), and such that

$$\left(\lim_{\Omega\to 0}(\Omega^{-1}f)\right)\circ\sigma_t^{-1}=c\,. \tag{4.54}$$

Then we define $(\tilde{\pi}_s,\tilde{\phi}_s,c_s)$ by calculating the initial data $(\tilde{\pi}_s,\tilde{\phi}_s)$ corresponding to f on $\mathscr{S}_s$, while c_s is obtained from the equation

$$c_s=\left(\lim_{\Omega\to 0}(\Omega^{-1}f)\right)\circ\sigma_s^{-1}. \tag{4.55}$$

Finally, we set

$$T_{s,t}((\tilde{\pi},\tilde{\phi},c))\equiv(\tilde{\pi}_s,\tilde{\phi}_s,c_s)\,.$$

We note that with this definition we have $(\tilde{\pi}_t,\tilde{\phi}_t,c_t)=(\tilde{\pi},\tilde{\phi},c)$, hence $T_{t,t}=\mathrm{id}$, as expected. During such an evolution the hyperboloid part of Σ moves along the flow of the vector field $\partial/\partial\omega$, as before. Consequently, the hyperboloid part of the Hamiltonian is given by the previous formula (4.32). To calculate the remaining, $\mathscr{I}^+$ part of the Hamiltonian (3.33), observe first that $\mathscr{L}_\tau$ in (3.33) vanishes because the vector field $\partial/\partial\omega$ is tangent to $\mathscr{I}^+$, we thus need to calculate $\partial c_\tau/\partial\tau$. Now, formulae (4.54) and (4.55) imply for $\tilde{\omega}\in[-1,t]$:

$$c_t\left(\sigma_t^{-1}(\tilde{\omega})\right)=c_s\left(\sigma_s^{-1}(\tilde{\omega})\right)\,.$$

Setting $s=t+\epsilon$ and denoting $\omega:=\sigma_{t+\epsilon}^{-1}(\tilde{\omega})\in[-1,0]$, we obtain:

$$c_{t+\epsilon}(\omega)=c_t\left(\sigma_t^{-1}\circ\sigma_{t+\epsilon}(\omega)\right)\,. \tag{4.56}$$

The derivation of the above equation with respect to ϵ at $\epsilon=0$ gives

$$\frac{\partial c_t}{\partial t}(\omega)=\pounds_Z c_t(\omega)\,, \tag{4.57}$$

where Z denotes the following (time dependent) vector field on $\mathbb{R}\times\mathscr{I}^+_{[-1,0]}$:

$$Z(t,\omega)\equiv Z_t(\omega)\frac{\partial}{\partial\omega}=\left.\frac{\partial\left(\sigma_t^{-1}\circ\sigma_{t+\epsilon}\right)(\omega)}{\partial\epsilon}\right|_{\epsilon=0}\frac{\partial}{\partial\omega}\,. \tag{4.58}$$

Inserting this into (3.33) yields

$$\mathscr{H}_{\mathscr{I}^+_{[-1,0]}}(c,\tau) = \int_{[-1,0]} d\omega \int_{\times S^2} \sin\theta \left(\frac{\partial c}{\partial \omega}\right)^2 Z_\tau(\omega)\, d\theta\, d\varphi\,. \tag{4.59}$$

Hence

$$\begin{aligned} H(\frac{\partial}{\partial\omega}, \mathscr{S}_0 \cup \mathscr{I}^+_{[-1,0]}, \cdot) &= \int_0^1 d\rho \int_{S^2} H^\omega(X,p)\, d\theta\, d\varphi \\ &+ \int_{[-1,0]} d\omega \int_{\times S^2} \sin\theta \left(\frac{\partial c}{\partial \omega}\right)^2 Z_\tau(\omega)\, d\theta\, d\varphi\,, \end{aligned} \tag{4.60}$$

with H^ω given by equation (4.30). Moreover, the boundary terms in (4.39) are multiplied by the value $Z_t(0) \equiv 1$ (the first one) and by $Z_t(-1) \equiv 0$ (the last one). Hence, gluing together the hyperboloid with the $\mathscr{I}^+$ part gives us the formula analogous to (4.41) with no boundary term.

In full analogy with formula (4.40), the fact that $\mathscr{H}_{\mathscr{I}^+_{[-1,0]}}(c,\tau)$ given by equation (4.59) is a Hamiltonian for the dynamics (4.57) may be checked directly by inspection:

$$\begin{aligned} \delta\left[\left(\frac{\partial c}{\partial\omega}\right)^2 Z_t(\omega)\right] &= 2Z_t(\omega)\frac{\partial c}{\partial\omega}\frac{\partial\delta c}{\partial\omega} = Z_t(\omega)\frac{\partial c}{\partial\omega}\frac{\partial\delta c}{\partial\omega} \\ &+ \frac{\partial}{\partial\omega}\left(Z_t(\omega)\frac{\partial c}{\partial\omega}\delta c\right) - \frac{\partial}{\partial\omega}\left(Z_t(\omega)\frac{\partial c}{\partial\omega}\right)\delta c \\ &= \frac{\partial\delta c}{\partial\omega}\pounds_{Z_t}c - \frac{\partial \pounds_{Z_t}c}{\partial\omega}\delta c + \frac{\partial}{\partial\omega}\left(Z_t(\omega)\frac{\partial c}{\partial\omega}\delta c\right), \end{aligned} \tag{4.61}$$

which gives the claimed result when integrated over $\mathscr{I}^+_{[-1,0]}$.

4.6 The preferred Hamiltonian role of the Trautman–Bondi energy for scalar fields

In physics one would like to be able to determine the maximal amount of energy which can be extracted out of a given field configuration. Hamiltonian frameworks have often been used as devices for producing candidates for the appropriate energy expressions. Now, a Hamiltonian approach can only give a convincing result if the expressions so obtained are unique. As already pointed out at the end of Sect. 2.1, Hamiltonians are indeed unique (modulo an additive constant), when the form Ω and the vector field $\mathfrak{X}_t$ of (2.7) are prescribed, and when the phase space is path connected. Now, we have just seen that for a massless scalar field in a radiation regime several phase spaces can be constructed. In each of those spaces the hyperboloids are moved by rigid time translations in Minkowski space, but the numerical values of the Hamiltonians differ (in so far as they can be compared at all, which in itself points out to yet another uniqueness problem). Now, in all the Hamiltonian systems associated with the time translations that are constructed in Chap. 4, one of the terms, that always appears in the Hamiltonian, is the Trautman–Bondi energy of the scalar field on a hyperboloid:

$$E_{TB}(\mathscr{S}_\tau) = \int_{\mathscr{S}_\tau} \mathscr{T}^\mu{}_\nu X^\nu dS_\mu \,, \tag{4.62}$$
$$X^\nu \partial_\nu = \partial_\omega \,, \qquad \mathscr{T}^\mu{}_\nu = \nabla^\mu f \nabla_\nu f - \tfrac{1}{2}\nabla^\alpha f \nabla_\alpha f \delta^\mu_\nu \,,$$

dS_μ as in Equation (3.38). In this section we wish to analyze in detail in which sense the "energy" (4.62) plays a preferred role in our considerations in this paper. Consider, first, the phase space $\mathscr{P}_{(-\infty,0]}$ of Chap. 4; in this phase space the Hamiltonian H is the sum of the Trautman–Bondi energy of $\mathscr{S}_0$ and of a term which can be interpreted as the flux of the energy radiated through $\mathscr{I}^+$. One expects that for a large class of fields H will be equal to the ADM energy (2.3) of the field, calculated at any hypersurface of constant Minkowskian time t. From this point of view the separation of H into the Trautman–Bondi energy (4.62) and a flux term might be thought as a convenient way of calculating the ADM energy. However, one can write a lot of other splittings the ADM energy into "a $\mathscr{S}_0$ contribution and a $\mathscr{I}^+$ contribution", and the Hamiltonian approach does not distinguish between those. In any case the phase space $\mathscr{P}_{(-\infty,0]}$ contains only fields with very special behaviour for large early times on $\mathscr{I}^+$: this does not seem to be a natural condition when talking about the energy of a field configuration on a given hyperboloid, say $\mathscr{S}_0$.

Naively, the situation seems to be better in $\mathscr{P}_{[-1,0]}$ where, as already mentioned in the paragraph following equation (4.50), the numerical value of the Hamiltonian equals $E_{TB}(\mathscr{S}_{-1})$. However, consider any functional $\widehat{H}$ of the form

$$\widehat{H} = E_{TB}(\mathscr{S}_{-1}) + \int_{S_{-1}} F\left[\chi|_{S_{-1}}, \frac{\partial\chi}{\partial\omega}\Big|_{S_{-1}}, \ldots, \frac{\partial^k\chi}{\partial\omega^k}\Big|_{S_{-1}}\right] + a \,, \tag{4.63}$$

where a is a constant, and F is a density depending differentiably upon the arguments indicated together with a finite number of angular derivatives thereof, which satisfies

$$\int_{S_{-1}} F\left[\chi|_{S_{-1}}, 0, \frac{\partial^2\chi}{\partial\omega^2}\Big|_{S_{-1}}, \ldots, \frac{\partial^k\chi}{\partial\omega^k}\Big|_{S_{-1}}\right] = 0 \,. \tag{4.64}$$

From equation (4.35) we have:

$$\begin{aligned} -\delta\widehat{H} = &\int_{\mathscr{S}_{-1}} \left(\pounds_X \tilde{p}\, \delta\tilde{f} - \pounds_X \tilde{f} \delta\tilde{p}\right) \\ &- \int_{S_{-1}} \left\{ \left(\frac{\delta F}{\delta\chi} - \dot{\chi}_0\right)\delta\chi + \frac{\delta F}{\delta\chi^{(1)}}\delta\chi^{(1)} + \ldots + \frac{\delta F}{\delta\chi^{(k)}}\delta\chi^{(k)} \right\}, \end{aligned} \tag{4.65}$$

$\chi^{(k)} \equiv \partial^k\chi/\partial\omega^k$. For fields in $\mathscr{P}_{[-1,0]}$ the boundary integral vanishes by equation (4.64) so that $\widehat{H}$ also is a Hamiltonian in $\mathscr{P}_{[-1,0]}$. The constant α in (4.63) plays a trivial role, and can be gotten rid of by requiring that we consider only those functionals, which vanish on the configuration $f \equiv 0$. Henceforth we shall thus assume

$$\alpha = 0.$$

We do not know any natural "symplectic" prescription[4], which would allow us to get rid of the remaining freedom in H. Let us, however, point out that the Trautman–Bondi energy defined by equation (4.62) is the only Hamiltonian $H(\mathscr{S}, p)$ in $\mathscr{P}_{[-1,0]}$ associated with ω-translations, which is in the class (4.63), which vanishes on the trivial configuration ($\tilde{p} = 0, \tilde{f} = 0, \chi = 0$), and which satisfies

$$\forall\, \tau_1 \geq -1 \qquad H(\mathscr{S}_{\tau_1}, p_{\tau_1}) \leq H(\mathscr{S}_{-1}, p_{-1})$$

for all p_{-1}, where p_{τ_1} is obtained by evolving p_{-1} to the future. This follows from the fact that H belongs to the class of functionals considered in equation (15) of [42], and the assertion can be established by repeating the proof of Theorem 1 of that reference[5].

Let us finally turn our attention to the phase space $\widehat{\mathscr{P}}_{[-1,0]}$ of Section 4.5. In that case the Hamiltonian (4.60) differs from the Trautman–Bondi energy by the somewhat strange looking term (4.59). Now, a simple calculation using (4.56) with $t + \epsilon = \tau$ gives

$$\mathscr{H}_{\mathscr{I}^+_{[-1,0]}}(c_\tau, \tau) = \int_{[-1,\tau]} d\omega \int_{S^2} \sin\theta \left(\frac{\partial c_0}{\partial\omega}\right)^2 \frac{\partial\sigma_\tau}{\partial\tau} \circ \sigma_\tau^{-1}\, d\theta\, d\varphi\,. \tag{4.66}$$

It is convenient to restrict the choice of the maps σ_τ by the requirement that there exists a constant C such that we have

$$\sup_{\tau\in(-1,0]} \left|\frac{\partial\sigma_\tau}{\partial\tau}\right| \leq C\,. \tag{4.67}$$

(An example of a family of maps satisfying this requirement, as well as all the previous ones, is given by the map $\sigma_\tau(\omega,\theta,\varphi) = ((\tau+1)(\omega+1)-1,\theta,\varphi)$.) For maps satisfying (4.67) one obtains from equation (4.66)

[4] We note that a recent proposal of Wald and Zoupas [126] how to define energy, momentum, *etc.*, at null infinity does not apply to the scalar field case, since this is not a generally covariant field theory in the sense of [126], though presumably the approach of [126] could be reformulated in a way which would cover the scalar fields in Minkowski space-time. In any case, several conditions are imposed in [126], which are not natural within a symplectic framework.

[5] Actually, in the context of this work there is no dependence upon the function d which occurs in equation (15) of [42], which simplifies somewhat the proof given in [42].

$$\lim_{\tau\to-1} \mathscr{H}_{\mathscr{I}^+_{[-1,0]}}(c_\tau,\tau) = 0 \,,$$

hence the limit as τ tends to -1 of the Hamiltonian (4.60) in $\widehat{\mathscr{P}}_{[-1,0]}$ equals the Trautman–Bondi mass (4.62).

4.7 The Poincaré group

So far we have shown how to construct Hamiltonians associated to time translations in Minkowski space-time. It is of interest to study motions of hypersurfaces in M by other isometries of $\mathbb{R}^{1,3}$. In this section we will verify that those motions preserve the phase spaces $\mathscr{P}_{(-\infty,0]}$ and $\mathscr{P}_{[-1,0]}$ constructed in Sects. 4.3 and 4.4, and show existence of the associated Hamiltonians. An analysis similar to that of Sect. 4.5 can be performed to obtain appropriate Hamiltonians on the phase space $\widehat{\mathscr{P}}_{[-1,0]}$: one simply needs to construct diffeomorphisms from $\mathscr{I}^+_{[-1,0]}$ to $\mathscr{I}^+_{[-1,\tau]}$, which interpolate between the identity at S_{-1} and the action of the Poincaré transformation under consideration at S_τ, the details are left to the reader.

Consider, then, the remaining nine generators of the Poincaré group. With respect to a fixed rest frame, they naturally split into three three-dimensional subspaces: space translations, rotations and boosts. Each of these subspaces may be given in terms of a "dipole-like" function v on S^2, *i.e.*, a function of angles which belongs to the first non-trivial (three-dimensional) eigenspace of the Laplacian on S^2. It is convenient to view S^2 as an embedded submanifold of $\mathbb{R}^3$, then this eigenspace is spanned by linear functions on $\mathbb{R}^3$

$$v(x^A) = v^i \frac{x_i}{r} \,,$$

and there is a one-to-one correspondence between functions v and vectors $(v^i) \in \mathbb{R}^3$. In spherical coordinates, the space of all v's is spanned by the functions $(\sin\theta\cos\varphi, \sin\theta\sin\varphi, \cos\theta)$.

It is easy to check that the following formulae hold in coordinates (t, x^A, r):

$$\begin{aligned}
X_{trans}(v) &= v\frac{\partial}{\partial r} + \frac{1}{r}\breve{h}^{AB}\partial_A v \frac{\partial}{\partial x^B} \,, \\
X_{rot}(v) &= \epsilon^{AB}\partial_B v \frac{\partial}{\partial x^A} \,, \\
X_{boost}(v) &= tX_{trans}(v) + rv\frac{\partial}{\partial t} \\
&= v\left(t\frac{\partial}{\partial r} + r\frac{\partial}{\partial t}\right) + \frac{t}{r}\breve{h}^{AB}\partial_A v \frac{\partial}{\partial x^B} \,,
\end{aligned}$$

where $(x^A) = (y^A)$ are spherical coordinates on S^2, $\breve{h}^{AB}$ is as in (4.19), and the tensor

$$\epsilon^{AB} := \frac{1}{\sqrt{\det h}}\{A,B\}$$

is defined by the Levi–Civita skew-symmetric symbol (tensor density) $\{A,B\}$, where $\{1,2\} = -\{2,1\} = 1$ and $\{1,1\} = \{2,2\} = 0$. As an example take $v(x^A) = \cos\theta$. The above formulae give in this case:

1. a space translation in the direction of the z-axis:

$$X_{trans}(\cos\theta) = \cos\theta\frac{\partial}{\partial r} - \frac{1}{r}\sin\theta\frac{\partial}{\partial\theta} ,$$

2. a rotation around the z-axis:

$$X_{rot}(\cos\theta) = \frac{\partial}{\partial\varphi} ,$$

3. as well as a boost in the direction of the z-axis:

$$X_{boost}(\cos\theta) = \cos\theta\left(t\frac{\partial}{\partial r} + r\frac{\partial}{\partial t}\right) - \frac{t}{r}\sin\theta\frac{\partial}{\partial\theta} .$$

More generally, if $J^{\mu\nu}$ is a constant skewsymmetric tensor in Minkowski vector space, and if in Minkowskian coordinates x^μ we set

$$X = \eta_{\mu\sigma}J^{\mu\nu}x^\sigma\partial_\nu ,$$

then the part X^A of X tangent to a unit sphere reads

$$X^A = h^{AB}\partial_B\left(J^0{}_l\frac{x^l}{r}\right) + \varepsilon^{AB}\partial_B\left(J^{kl}\epsilon_{klm}\frac{x^m}{r}\right) . \tag{4.68}$$

Passing to the coordinates $(y^\mu) = (\omega, y^A, \rho)$ with $(y^A) = (x^A)$,

$$\omega = t - \sqrt{1+r^2} , \qquad \rho = \sqrt{1+\frac{1}{r^2}} - \frac{1}{r} ,$$

we obtain $\partial/\partial y^A = \partial/\partial x^A$ and

$$\frac{\partial}{\partial t} = \frac{\partial}{\partial\omega} ,$$
$$\frac{\partial}{\partial r} = -\frac{2\rho}{1+\rho^2}\frac{\partial}{\partial\omega} + \frac{\left(1-\rho^2\right)^2}{2\left(1+\rho^2\right)}\frac{\partial}{\partial\rho} .$$

Inserting these expressions into the definition of the corresponding Killing vectors, we obtain:

$$X_{trans}(v) = v\left(-\frac{2\rho}{1+\rho^2}\frac{\partial}{\partial\omega} + \frac{\left(1-\rho^2\right)^2}{2\left(1+\rho^2\right)}\frac{\partial}{\partial\rho}\right) + \frac{1-\rho^2}{2\rho}\breve{h}^{AB}\partial_A v\frac{\partial}{\partial y^B}\,, \tag{4.69}$$

$$X_{rot}(v) = \epsilon^{AB}\partial_B v\frac{\partial}{\partial y^A}\,, \tag{4.70}$$

$$X_{boost}(v) = v\left\{-\frac{2\omega\rho}{1+\rho^2}\frac{\partial}{\partial\omega} + \frac{1-\rho^2}{2}\left[\frac{\omega(1-\rho^2)}{1+\rho^2}+1\right]\frac{\partial}{\partial\rho}\right\} + \frac{\omega(1-\rho^2)+1+\rho^2}{2\rho}\breve{h}^{AB}\partial_A v\frac{\partial}{\partial y^B}\,. \tag{4.71}$$

We see that these fields are manifestly extendable to $\widetilde{M}$, with the corresponding extensions tangent to $\mathscr{I}^+$ (one can also obtain this result on general grounds by conformal covariance of the conformal Killing equations). This implies that the corresponding flows act smoothly on $\widetilde{M}$, mapping $\mathscr{I}^+$ to itself. The obvious equivalent of the prescription given in Sect. 4.3 allows us to define dynamical trajectories through points $p \in \mathscr{P}_{(-\infty,0]}$ associated to motions of $\mathscr{S}_0 \cup \mathscr{I}^+_{(-\infty,0]}$ along the one-parameter groups generated by those vector fields. It then immediately follows from the definition of the phase space $\mathscr{P}_{(-\infty,0]}$ that those trajectories remain in $\mathscr{P}_{(-\infty,0]}$, and lead thus to a dynamical system on $\mathscr{P}_{(-\infty,0]}$. Similarly, every one-parameter subgroup of the Poincaré group defines a *local* dynamical system on $\mathscr{P}_{[-1,0]}$.

The Hamiltonians corresponding to fields (4.69)–(4.71) are: momenta, angular momenta and boost generators. They are — as usual — obtained by integrating the quantity $H^\mu(X,p) = p^\mu \pounds_X f - X^\mu L$ (with X being one of the above generators) over the surface $\mathscr{S}_\tau \cup \mathscr{I}^+_{(\infty,\tau]}$. Actually, the only non-vanishing contribution to the integrals comes from H^ω on $\mathscr{S}_\tau = \{\omega = \tau\}$ and from H^ρ on $\mathscr{I}^+_{(\infty,\tau]} \subset \{\rho = 1\}$. We are going to prove in the sequel that both integrands extend continuously to $\widetilde{M}$, which implies that the integrals in question are finite. For this purpose let us, at first, analyze the asymptotic behaviour of L. We have:

$$\begin{aligned} L &= -\frac{1}{2}\sqrt{-\eta}\;\eta^{\mu\nu}\partial_\mu f\partial_\nu f = -\frac{1}{2}\sqrt{-\tilde{g}}\;\tilde{g}^{\mu\nu}\frac{1}{\Omega^2}\partial_\mu(\Omega\tilde{f})\partial_\nu(\Omega\tilde{f}) \\ &= -\frac{1}{2}\sqrt{-\tilde{g}}\;\tilde{g}^{\mu\nu}\partial_\mu\tilde{f}\partial_\nu\tilde{f} - \sqrt{-\tilde{g}}\;\tilde{g}^{\mu\nu}\frac{1}{\Omega}\tilde{f}\partial_\mu\tilde{f}\partial_\nu\Omega \\ &\quad -\frac{1}{2}\sqrt{-\tilde{g}}\;\tilde{g}^{\mu\nu}\tilde{f}^2\frac{1}{\Omega^2}(\partial_\mu\Omega)(\partial_\nu\Omega)\,. \end{aligned} \tag{4.72}$$

(Recall that the metric $\tilde{g}^{\mu\nu}$ has been defined in equation (4.14).) The first term is manifestly extendable to $\widetilde{M}$. Further, equation (4.16) implies that $\partial_\nu\Omega = -\rho\delta^\rho_\nu$. Inserting this into the last term, we obtain

$$-\frac{1}{2}\sqrt{-\tilde{g}}\;\tilde{g}^{\mu\nu}\tilde{f}^2\frac{1}{\Omega^2}(\partial_\mu\Omega)(\partial_\nu\Omega) = -\frac{1}{2}\sqrt{-\tilde{g}}\;\frac{\tilde{g}^{\rho\rho}}{\Omega^2}\rho^2\tilde{f}^2\,,$$

which is again extendable to $\widetilde{M}$ because, due to (4.15), the component $\tilde{g}^{\rho\rho}$ behaves like Ω^2 on $\mathscr{I}^+$. The only singular term in the Lagrangian comes thus from the second term in (4.72):

$$\begin{aligned}-\sqrt{-\tilde{g}}\,\tilde{g}^{\mu\nu}\frac{1}{\Omega}\tilde{f}\partial_\mu\tilde{f}\partial_\nu\Omega &= \sqrt{-\tilde{g}}\,\rho\frac{\tilde{g}^{\mu\rho}}{\Omega}\tilde{f}\partial_\mu\tilde{f}\\ &= \sqrt{-\tilde{g}}\,\rho\frac{\tilde{g}^{\rho\rho}}{\Omega}\tilde{f}\partial_\rho\tilde{f} + \sqrt{-\tilde{g}}\,\rho\frac{\tilde{g}^{\omega\rho}}{\Omega}\tilde{f}\partial_\omega\tilde{f}\,.\end{aligned}$$

Repeating the previous argument, one finds that only the last term is singular. Hence, we may rewrite the Lagrangian in the following way:

$$L = L_{regular} + \sqrt{-\tilde{g}}\,\rho\frac{\tilde{g}^{\omega\rho}}{\Omega}\tilde{f}\partial_\omega\tilde{f}\,, \tag{4.73}$$

where $L_{regular}$ is regular on the entire $\widetilde{M}$. Moreover, we have

$$\begin{aligned}p^\mu \pounds_X f &= p^\mu \pounds_X(\Omega\tilde{f}) = \Omega p^\mu \pounds_X\tilde{f} + p^\mu\tilde{f}\pounds_X\Omega = \Omega p^\mu\left(\pounds_X\tilde{f} - \rho\frac{X^\rho}{\Omega}\tilde{f}\right)\\ &= \left(\tilde{p}^\mu - \sqrt{-\tilde{g}}\,\tilde{g}^{\mu\nu}\frac{1}{\Omega}\tilde{f}\partial_\nu\Omega\right)\left(\pounds_X\tilde{f} - \rho\frac{X^\rho}{\Omega}\tilde{f}\right)\\ &= \left(\tilde{p}^\mu + \sqrt{-\tilde{g}}\,\rho\frac{\tilde{g}^{\mu\rho}}{\Omega}\tilde{f}\right)\left(X^\omega\partial_\omega\tilde{f} + X^\rho\partial_\rho\tilde{f} + X^A\partial_A\tilde{f} - \rho\frac{X^\rho}{\Omega}\tilde{f}\right)\,.\end{aligned}$$

For fields X given by formulae (4.69)–(4.71) all the terms in the second bracket are regular on $\widetilde{M}$, because X^ρ vanishes at least as fast as Ω on $\mathscr{I}^+$. Next, equations (4.14) and (4.18) give $\tilde{g}^{\mu\rho} = \delta^\mu_\omega$ so that the potentially dangerous term $\tilde{g}^{\mu\rho}/\Omega$ above matters only for $\mu = \omega$. We claim that this term integrated over the sphere S^2 gives a vanishing contribution in the limit $\rho \to 1$:

$$\begin{aligned}&\lim_{\rho\to 1}\int_{S^2}\sqrt{-\tilde{g}}\,\frac{\rho\tilde{f}}{\Omega}\left(X^A\partial_A\tilde{f} - \rho\frac{X^\rho}{\Omega}\tilde{f}\right)d\theta\, d\varphi =\\ &\lim_{\rho\to 1}\int_{S^2}\sqrt{-\tilde{g}}\,\frac{\rho}{\Omega}\left(\frac{1}{2}X^A\partial_A(\tilde{f}^2) - \rho\frac{X^\rho}{\Omega}\tilde{f}^2\right)d\theta\, d\varphi = 0\end{aligned} \tag{4.74}$$

To prove this statement we integrate by parts the first term; as a result we obtain a term containing the two-dimensional divergence of the field X^A. In case of the field X_{rot} the latter vanishes, with the component X^ρ_{rot} vanishing as well. This proves the statement. In case of X_{trans} and X_{boost} the argument is slightly more involved: the divergence of X^A equals the two-dimensional Laplacian of the function v. As v is an eigenfunction with eigenvalue equal to -2, the term containing the divergence of X^A reproduces the second term modulo a factor which tends to -1 on $\mathscr{I}^+$. We conclude that $p^\mu\pounds_X f$ may also be decomposed into a regular and a singular part as follows:

$$p^\mu\pounds_X f = (p^\mu\pounds_X f)_{regular} + X^\omega\sqrt{-\tilde{g}}\,\rho\frac{\tilde{g}^{\mu\rho}}{\Omega}\tilde{f}\partial_\omega\tilde{f}\,. \tag{4.75}$$

Strictly speaking, the term $(p^\mu \pounds_X f)_{regular}$ is not regular by itself, but gives a finite contribution after having been integrated upon, as in equation (4.74). Combining equations (4.73) and (4.75) we are led to the following conclusions:

- H^ρ is regular on $\widetilde{M}$ because $\tilde{g}^{\rho\rho}$ in (4.75) vanishes like Ω^2, whereas (4.73) becomes regular when multiplied by X^ρ, vanishing on $\mathscr{I}^+$ at least like Ω;
- H^ω is regular because the singular term in (4.75) cancels with the singular part of (4.73) multiplied by X^ω.

We have thus verified that the integrals $H(X, \mathscr{S}_0 \cup \mathscr{I}^+_{(-\infty,0]}, p)$ for $p \in \mathscr{P}_{(-\infty,0]}$, and $H(X, \mathscr{S}_0 \cup \mathscr{I}^+_{[-1,0]}, p)$ for $p \in \mathscr{P}_{[-1,0]}$, given by equation (3.41), are convergent. If, finally, we consider the variations of those integrals, the unwanted boundary terms in equation (3.45) vanish precisely for the same reason as they did when X was a time translation. This finishes the proof, that every one-parameter subgroup of the Poincaré group generates a Hamiltonian dynamical system in $\mathscr{P}_{(-\infty,0]}$ and $\mathscr{P}_{[-1,0]}$.

4.8 "Supertranslated" hyperbolae

In the previous sections we have, for the sake of simplicity, restricted ourselves to hypersurfaces $\mathscr{S}$ which are hyperbolae in Minkowski space-time. In some situations it might be of interest to consider the class of hypersurfaces, given in a neighbourhood of $\rho = 1$ by the equation

$$\mathscr{S} = \{\omega = \alpha(\rho, \theta, \varphi)\} \ , \tag{4.76}$$

for some smooth function α; we call such hypersurfaces *supertranslated hyperbolae*. If α is a constant, one recovers the standard hyperbolae; on the other hand, a translated hyperbola is described by a function α which at $\rho = 1$ will be a linear combination of $\ell = 0$ and $\ell = 1$ spherical harmonics. We shall not make an exhaustive analysis of the dynamics of such hypersurfaces, which can be done by a rather direct repetition of the arguments presented in the previous section, and we shall only analyze the question of convergence of the putative Hamiltonians (4.13).

Consider any vector field X, which is smooth up-to-boundary on the future conformally completed Minkowski space-time $\widetilde{M}$, and tangent to $\mathscr{I}^+$. We note that in particular the generators of the Poincaré group do have this property, which follows either by general considerations concerning isometries, or by direct examination of equations (4.69)–(4.71). The need for X to be tangent to $\mathscr{I}^+$ follows from the requirement that points in the physical space-time M remain in M when moved with the flow of X. For such vectors $\Omega^{-1}X^\rho$ is a smooth up-to-boundary function on $\widetilde{M}$, hence the same holds for $\pounds_X(\ln \Omega)$:

$$\pounds_X(\ln \Omega) = \Omega^{-1}X^\rho \partial_\rho \Omega \ .$$

We have the identity

$$\begin{aligned} H^\mu &:= p^\mu \pounds_X f - X^\mu L \\ &= \tilde{p}^\mu X^\lambda \partial_\lambda \tilde{f} - X^\mu \tilde{L} + \tilde{p}^\mu \tilde{f} \pounds_X (\ln \Omega) + \frac{1}{2} \Omega^2 \tilde{f}^2 \pounds_X Z^\mu \\ &\quad + \frac{1}{2} \partial_\lambda \left[\Omega^2 \tilde{f}^2 (X^\mu Z^\lambda - Z^\mu X^\lambda) \right] , \end{aligned} \tag{4.77}$$

where

$$Z^\mu := \Omega^{-3} \sqrt{-\tilde{g}} \tilde{g}^{\lambda\mu} \partial_\mu \Omega \,,$$

and where $\tilde{L}$ is given by equation (4.27). We note that it follows from our previous considerations that the vector field

$$\Omega^2 \pounds_X Z^\lambda = \sqrt{-\tilde{g}} \tilde{g}^{\lambda\mu} \partial_\mu (\Omega^{-1} X^\rho \partial_\rho \Omega) = \sqrt{-\tilde{g}} \tilde{g}^{\lambda\mu} \partial_\mu (\pounds_X (\ln \Omega))$$

is again smooth up-to-boundary on $\widetilde{M}$. Further $Z^A = 0$, while $\Omega^3 Z^\omega$ and ΩZ^ρ smoothly extend to $\mathscr{I}^+$. For any vector density Y^μ and for hypersurfaces $\mathscr{S}$ described by equation (4.76) we have

$$\int_{\mathscr{S}} Y^\mu \, dS_\mu = \int (Y^\omega - Y^k \alpha_{,k}) \, d\rho \, d\theta \, d\varphi \tag{4.78}$$

(the last integral being performed on the range of the relevant variables in which equation (4.76) holds), which shows that all the terms at the right hand side of (4.77) except perhaps for the last one will give a convergent integral when integrated upon a hypersurface of the form (4.76). To analyze this last term we note that on $\mathscr{S}^\epsilon \equiv \mathscr{S} \cap \{\rho = 1 - \epsilon\}$ we have

$$\begin{aligned} \int_{\mathscr{S}^\epsilon} \partial_\lambda & \left[\Omega^2 \tilde{f}^2 (X^\mu Z^\lambda - Z^\mu X^\lambda) \right] dS_\mu = \\ &= \int_{\partial \mathscr{S}^\epsilon} \Omega^2 \tilde{f}^2 \left[(X^\rho Z^\omega - Z^\rho X^\omega) - (X^\rho Z^A - Z^\rho X^A) \partial_A \alpha \right] d\theta \, d\varphi \\ &= \int_{\partial \mathscr{S}^\epsilon} \Omega^2 \tilde{f}^2 \left[X^\rho Z^\omega - Z^\rho X^\omega + Z^\rho X^A \partial_A \alpha \right] d\theta \, d\varphi \,, \end{aligned}$$

where we have used $Z^A = 0$. Since $\Omega^{-1} X^\rho$, $\Omega^3 Z^\omega$ and ΩZ^ρ are finite at $\mathscr{I}^+$ one can safely pass to the limit $\epsilon \to 0$ to obtain a finite convergent contribution to $\int_{\mathscr{S}} H^\mu \, dS_\mu$ from the last term in (4.77) as well, for all vector fields as considered above, in particular for the generators of the Poincaré group. In the case of constant α this gives an alternative derivation of the corresponding analysis of Sect. 4.7. It should be emphasized that in the current setting in general the integrand of $\int_{\mathscr{S}} H^\mu \, dS_\mu$ is not integrable, so that $\int_{\mathscr{S}} H^\mu \, dS_\mu$ has to be defined by the limiting procedure described above.

Some of the "supertranslated hyperbolae" may be obtained (at least in a neighbourhood of $\mathscr{I}^+$) from the standard ones by the flow of vector fields called "supertranslations", which arise naturally when considering asymptotically flat radiating metrics (this is discussed in more detail in Sect. 6.2

below). Those vector fields are defined by the requirement that: 1) they extend smoothly to $\mathscr{I}^+$; 2) they preserve the Bondi form of the metric; and 3) $X|_{\mathscr{I}^+} = \lambda(\theta,\varphi)\partial_\omega$ for some smooth function λ on S^2. In Minkowski space-time, and away from $\rho = 0$, they are of the form [83, p. 724]

$$X = \left(\lambda + \Omega^2 \frac{\Delta_2 \lambda}{(1+\rho^2)(1+\rho)}\right)\frac{\partial}{\partial\omega} - \frac{\Omega}{\rho}\breve{h}^{AB}\partial_A\lambda\frac{\partial}{\partial y^B} + \Omega^2\frac{\Delta_2\lambda}{1+\rho^2}\frac{\partial}{\partial\rho}\,, \tag{4.79}$$

where Δ_2 is the Laplacian of the standard round metric $\breve{h}_{AB}$ on a two-dimensional unit sphere. Strictly speaking, the case where λ is a linear combination of $\ell = 0$ and $\ell = 1$ harmonics corresponds to translations, and should thus be excluded from this definition. It is, however, convenient to treat translations and supertranslations on the same footing, and we will often do so. Equation (4.79) shows that the supertranslations satisfy the requirements on X set forth above, and thus the corresponding integrals $\int_{\mathscr{S}} H^\mu \, dS_\mu$ are well defined. The collection of Poincaré transformations and supertranslations will be referred to as the set of *generators of the BMS group.*

Let us close this section by deriving the corner conditions which occur in the "supertranslated hyperbolae" case. First, the continuity condition (4.42) remains unchanged. Next, to analyse the differentiability condition (4.43), let $\tilde{p} \equiv \tilde{p}^{\hat{\omega}}$, where $\hat{\omega} \equiv \omega - \alpha$; we find

$$\begin{aligned}\tilde{p} = {} & \frac{2\rho^2\sin\theta}{(1+\rho^2)}\left[\frac{\partial\tilde{f}}{\partial\omega} + \rho\frac{\partial\tilde{f}}{\partial\rho}\right] - \alpha_{,\rho}\frac{\rho^2\sin\theta}{1+\rho^2}\left[2\rho\frac{\partial\tilde{f}}{\partial\omega} - \frac{(\rho^2-1)^2}{2}\frac{\partial\tilde{f}}{\partial\rho}\right] \\ & - \frac{1+\rho^2}{2}\sin\theta\breve{h}^{AB}\frac{\partial\tilde{f}}{\partial x^A}\alpha_{,B}\,.\end{aligned} \tag{4.80}$$

Let $\tilde{\phi}$, respectively $\tilde{\pi}$, be the pull-back to Σ of $\tilde{f}$, respectively $\tilde{p}$; it follows from the discussion in Chap. 3 that $\tilde{\pi}$ represents the momentum density on Σ associated with $\tilde{\phi}$; from equation (4.80) one finds that the differentiability condition (4.43) is replaced by

$$\left.\left(\frac{\tilde{\pi}}{\sin\theta} - \frac{\partial\tilde{\phi}}{\partial\rho} + \breve{h}^{AB}\frac{\partial\tilde{\phi}}{\partial x^A}\alpha_{,B}\right)\right|_{\rho=1} = (1-\alpha_{,\rho})\left.\frac{\partial c}{\partial\omega}\right|_{\omega=0}\,. \tag{4.81}$$

5. The energy of the gravitational field

5.1 Preliminaries

In this section we will consider the gravitational field on a space-time M, described by a metric tensor $g_{\mu\nu}$, of signature $(-1,+1,+1,+1)$, which satisfies the vacuum Einstein equations:

$$R_{\mu\nu}(g) = 0\,. \tag{5.1}$$

There exist several variational principles which produce those equations [26, 47, 54, 56, 57, 70, 76, 77, 86, 91, 103], as well as several canonical formulations of the associated theory [8, 13, 19, 21, 71, 73, 102, 107]. Here, we are going to follow a generalization of the background field formulation used in [38], based on a first order variational principle "covariations" with a background connection[1]. Recall that in the original variational principle of Einstein [57] one removes from the Hilbert Lagrangian [76] a coordinate-dependent divergence, obtaining thus a first order variational principle for the metric. In [38] the Hilbert Lagrangian was modified by removing a coordinate-independent divergence, which, however, did depend upon a background metric. Let us start by reviewing the approach of [38]: consider the Ricci tensor,

$$R_{\mu\nu} = \partial_\alpha \left[\Gamma^\alpha_{\mu\nu} - \delta^\alpha_{(\mu}\Gamma^\kappa_{\nu)\kappa} \right] - \left[\Gamma^\alpha_{\sigma\mu}\Gamma^\sigma_{\alpha\nu} - \Gamma^\alpha_{\mu\nu}\Gamma^\sigma_{\alpha\sigma} \right] , \tag{5.2}$$

where the Γ's are the Christoffel symbols of g. Contracting $R_{\mu\nu}$ with the contravariant density of metric,

$$\mathfrak{g}^{\mu\nu} := \frac{1}{16\pi}\sqrt{-\det g}\; g^{\mu\nu}\,, \tag{5.3}$$

one obtains the following expression for the Hilbert Lagrangian density:

$$\begin{aligned} \tilde{L} &= \frac{1}{16\pi}\sqrt{-\det g R} = \mathfrak{g}^{\mu\nu} R_{\mu\nu} \\ &= \partial_\alpha \left[\mathfrak{g}^{\mu\nu} \left(\Gamma^\alpha_{\mu\nu} - \delta^\alpha_{(\mu}\Gamma^\kappa_{\nu)\kappa} \right) \right] + \mathfrak{g}^{\mu\nu} \left[\Gamma^\alpha_{\sigma\mu}\Gamma^\sigma_{\alpha\nu} - \Gamma^\alpha_{\mu\nu}\Gamma^\sigma_{\alpha\sigma} \right] . \end{aligned} \tag{5.4}$$

[1] The idea of using a background connection instead of a background metric has been advocated in [115]. However, the framework we use here is closer in spirit to the one in [38].

Here we have used the metricity condition of Γ, which is equivalent to the following identity:

$$\mathfrak{g}^{\mu\nu}{}_{,\alpha} := \partial_\alpha \mathfrak{g}^{\mu\nu} = \mathfrak{g}^{\mu\nu} \Gamma^\sigma_{\alpha\sigma} - \mathfrak{g}^{\mu\sigma} \Gamma^\nu_{\sigma\alpha} - \mathfrak{g}^{\nu\sigma} \Gamma^\mu_{\sigma\alpha} \, . \tag{5.5}$$

Suppose now, that $B^\alpha_{\sigma\mu}$ is another symmetric connection in M, which will be used as a background (or reference) connection. Denote by $r_{\mu\nu}$ its Ricci tensor. From the metricity condition (5.5) we similarly obtain

$$\begin{aligned} \mathfrak{g}^{\mu\nu} r_{\mu\nu} = \partial_\alpha \left[\mathfrak{g}^{\mu\nu} \left(B^\alpha_{\mu\nu} - \delta^\alpha_{(\mu} B^\kappa_{\nu)\kappa} \right) \right] - \mathfrak{g}^{\mu\nu} \left[B^\alpha_{\sigma\mu} B^\sigma_{\alpha\nu} - B^\alpha_{\mu\nu} B^\sigma_{\alpha\sigma} \right] \\ + \mathfrak{g}^{\mu\nu} \left[\Gamma^\alpha_{\sigma\mu} B^\sigma_{\alpha\nu} + B^\alpha_{\sigma\mu} \Gamma^\sigma_{\alpha\nu} - \Gamma^\alpha_{\mu\nu} B^\sigma_{\alpha\sigma} - B^\alpha_{\mu\nu} \Gamma^\sigma_{\alpha\sigma} \right] . \end{aligned} \tag{5.6}$$

It is useful to introduce the tensor field

$$p^\alpha_{\mu\nu} := \left(B^\alpha_{\mu\nu} - \delta^\alpha_{(\mu} B^\kappa_{\nu)\kappa} \right) - \left(\Gamma^\alpha_{\mu\nu} - \delta^\alpha_{(\mu} \Gamma^\kappa_{\nu)\kappa} \right) . \tag{5.7}$$

Once the reference connection $B^\alpha_{\mu\nu}$ is given, the tensor $p^\alpha_{\mu\nu}$ encodes the entire information about the connection $\Gamma^\alpha_{\mu\nu}$:

$$\Gamma^\alpha_{\mu\nu} = B^\alpha_{\mu\nu} - p^\alpha_{\mu\nu} + \frac{2}{3} \delta^\alpha_{(\mu} p^\kappa_{\nu)\kappa} \, . \tag{5.8}$$

Subtracting equation (5.6) from (5.4), and using the definition of $p^\alpha_{\mu\nu}$, we arrive at the equation

$$\mathfrak{g}^{\mu\nu} R_{\mu\nu} = -\partial_\alpha \left(\mathfrak{g}^{\mu\nu} p^\alpha_{\mu\nu} \right) + L \, ,$$

where

$$L := \mathfrak{g}^{\mu\nu} \left[\left(\Gamma^\alpha_{\sigma\mu} - B^\alpha_{\sigma\mu} \right) \left(\Gamma^\sigma_{\alpha\nu} - B^\sigma_{\alpha\nu} \right) - \left(\Gamma^\alpha_{\mu\nu} - B^\alpha_{\mu\nu} \right) \left(\Gamma^\sigma_{\alpha\sigma} - B^\sigma_{\alpha\sigma} \right) + r_{\mu\nu} \right] .$$

This result may be used as follows: the quantity L differs by a total divergence from the Hilbert Lagrangian, and hence the associated variational principle leads to the same equations of motion. Further, the metricity condition (5.5) enables us to rewrite L in terms of the first derivatives of $\mathfrak{g}^{\mu\nu}$: indeed, replacing in (5.5) the partial derivatives $\mathfrak{g}^{\mu\nu}{}_{,\alpha}$ by the covariant derivatives $\mathfrak{g}^{\mu\nu}{}_{;\alpha}$, calculated with respect to the background connection B,

$$\mathfrak{g}^{\mu\nu}{}_{;\alpha} := \mathfrak{g}^{\mu\nu} \left(\Gamma^\sigma_{\alpha\sigma} - B^\sigma_{\alpha\sigma} \right) - \mathfrak{g}^{\mu\sigma} \left(\Gamma^\nu_{\sigma\alpha} - B^\nu_{\sigma\alpha} \right) - \mathfrak{g}^{\nu\sigma} \left(\Gamma^\mu_{\sigma\alpha} - B^\mu_{\sigma\alpha} \right) , \tag{5.9}$$

we may calculate $p^\alpha_{\mu\nu}$ in terms of the latter derivatives. The final result is:

$$\begin{aligned} p^\lambda_{\mu\nu} = \frac{1}{2} \mathfrak{g}_{\mu\alpha} \mathfrak{g}^{\lambda\alpha}{}_{;\nu} + \frac{1}{2} \mathfrak{g}_{\nu\alpha} \mathfrak{g}^{\lambda\alpha}{}_{;\mu} - \frac{1}{2} \mathfrak{g}^{\lambda\alpha} \mathfrak{g}_{\sigma\mu} \mathfrak{g}_{\rho\nu} \mathfrak{g}^{\sigma\rho}{}_{;\alpha} \\ + \frac{1}{4} \mathfrak{g}^{\lambda\alpha} \mathfrak{g}_{\mu\nu} \mathfrak{g}_{\sigma\rho} \mathfrak{g}^{\sigma\rho}{}_{;\alpha} \, , \end{aligned} \tag{5.10}$$

where by $\mathfrak{g}_{\mu\nu}$ we denote the matrix inverse to $\mathfrak{g}^{\mu\nu}$. Inserting these results into the definition of L, we obtain:

$$\begin{aligned} L = & \frac{1}{2}\mathfrak{g}_{\mu\alpha}\mathfrak{g}^{\mu\nu}{}_{;\lambda}\mathfrak{g}^{\lambda\alpha}{}_{;\nu} - \frac{1}{4}\mathfrak{g}^{\lambda\alpha}\mathfrak{g}_{\sigma\mu}\mathfrak{g}_{\rho\nu}\mathfrak{g}^{\mu\nu}{}_{;\lambda}\mathfrak{g}^{\sigma\rho}{}_{;\alpha} \\ & + \frac{1}{8}\mathfrak{g}^{\lambda\alpha}\mathfrak{g}_{\mu\nu}\mathfrak{g}^{\mu\nu}{}_{;\lambda}\mathfrak{g}_{\sigma\rho}\mathfrak{g}^{\sigma\rho}{}_{;\alpha} + \mathfrak{g}^{\mu\nu}r_{\mu\nu}\,. \end{aligned} \tag{5.11}$$

We note the identity

$$\frac{\partial L}{\partial \mathfrak{g}^{\mu\nu}{}_{,\lambda}} = \frac{\partial L}{\partial \mathfrak{g}^{\mu\nu}{}_{;\lambda}} = p^{\lambda}_{\mu\nu}\,, \tag{5.12}$$

which shows that the connection Γ — described by means of the tensor $p^{\lambda}_{\mu\nu}$ — is the momentum canonically conjugate to the contravariant tensor density $\mathfrak{g}^{\mu\nu}$; prescribing this last object is of course equivalent to prescribing the metric. From this point of view gravitational fields on a manifold M are sections of the bundle

$$\begin{array}{c} F = S_0T^2M \otimes \tilde{\Lambda}^4 M \\ \Big\downarrow {\scriptstyle pr_{F\to M}} \\ M \end{array} \tag{5.13}$$

where S_0T^2M denotes the bundle of non-degenerate symmetric contravariant tensors over M. Given a background symmetric connection B on M, we take L given by equation (5.11) as the Lagrangian for the theory. The canonical momentum $p^{\lambda}_{\mu\nu}$ is defined by equation (5.10). If $\mathscr{S}$ is any piecewise smooth hypersurface in M, and if $(\delta_a p^{\lambda}_{\mu\nu}, \delta_a\mathfrak{g}^{\alpha\beta})$, $a = 1, 2$, are two sections over $\mathscr{S}$ of the bundle of vertical vectors tangent to the space-time phase bundle (as defined in Sect. 3.2), following equation (3.37) we set

$$\Omega_{\mathscr{S}}((\delta_1 p^{\lambda}_{\mu\nu}, \delta_1\mathfrak{g}^{\alpha\beta}), (\delta_2 p^{\lambda}_{\mu\nu}, \delta_2\mathfrak{g}^{\alpha\beta})) = \int_{\mathscr{S}} (\delta_1 p^{\mu}_{\alpha\beta}\delta_2\mathfrak{g}^{\alpha\beta} - \delta_2 p^{\mu}_{\alpha\beta}\delta_1\mathfrak{g}^{\alpha\beta})\, dS_{\mu}\,. \tag{5.14}$$

As emphasized in Chap. 3, only such fields $(\delta_a p^{\lambda}_{\mu\nu}, \delta_a\mathfrak{g}^{\alpha\beta})$ are allowed in (5.14), which arise from some solutions of the vacuum Einstein equations.

Suppose that the map ψ of Chap. 3 arises from the flow of a vector field $X = X^{\mu}\partial_{\mu}$ defined on M. In this case Ψ can be chosen to be the canonical (Lie) lift of ψ to the tensor bundle F, hence the derivative $\mathfrak{D}_X$ of equation (3.10) is $\pounds_X$. According to equation (3.41), in this situation the Hamiltonian equals

$$H(X, \mathscr{S}) = \int_{\mathscr{S}} (p^{\mu}_{\alpha\beta}\pounds_X\mathfrak{g}^{\alpha\beta} - X^{\mu}L)\, dS_{\mu}\,. \tag{5.15}$$

(To shorten notation from now on we shall omit the dependence upon the phase space variable $p = (\mathfrak{g}^{\mu\nu}, p^{\lambda}{}_{\alpha\beta})$ and write $H(X, \mathscr{S})$ for $H(X, \mathscr{S}, p)$, *etc.*) We are going to analyse the above quantity without any assumptions on the vector field X and on the background connection B. Before we do this, let us remind results which apply to the case when B is the metric connection of a given background metric $b_{\mu\nu}$, and when X is a Killing vector field of $b_{\mu\nu}$.

Under those restrictions it was shown in [38] that the integrand in (5.15) is equal to the divergence of a "Freud-type superpotential" [58]:

$$H^\mu \equiv p^\mu_{\alpha\beta}\pounds_X \mathfrak{g}^{\alpha\beta} - X^\mu L = \partial_\alpha \mathbb{W}^{\mu\alpha} , \tag{5.16}$$

$$\mathbb{W}^{\nu\lambda} = \mathbb{W}^{\nu\lambda}{}_\beta X^\beta - \frac{1}{8\pi}\sqrt{|\det g_{\rho\sigma}|}g^{\alpha[\nu}\delta^{\lambda]}_\beta X^\beta{}_{;\alpha} , \tag{5.17}$$

$$\begin{aligned}\mathbb{W}^{\nu\lambda}{}_\beta &= \frac{2|\det b_{\mu\nu}|}{16\pi\sqrt{|\det g_{\rho\sigma}|}} g_{\beta\gamma}(e^2 g^{\gamma[\lambda}g^{\nu]\kappa})_{;\kappa} \\ &= 2\mathfrak{g}_{\beta\gamma}\left(\mathfrak{g}^{\gamma[\lambda}\mathfrak{g}^{\nu]\kappa}\right)_{;\kappa} \\ &= 2\mathfrak{g}^{\mu[\nu}p^{\lambda]}_{\mu\beta} - 2\delta^{[\nu}_\beta p^{\lambda]}_{\mu\sigma}\mathfrak{g}^{\mu\sigma} - \frac{2}{3}\mathfrak{g}^{\mu[\nu}\delta^{\lambda]}_\beta p^\sigma_{\mu\sigma} ,\end{aligned} \tag{5.18}$$

where a semi-column denotes the covariant derivative of the metric b, square brackets denote antisymmetrization (with a factor of 1/2 when two indices are involved), as before $\mathfrak{g}_{\beta\gamma} \equiv (\mathfrak{g}^{\alpha\sigma})^{-1} = 16\pi g_{\beta\gamma}/\sqrt{|\det g_{\rho\sigma}|}$, and

$$e \equiv \frac{\sqrt{|\det g_{\rho\sigma}|}}{\sqrt{|\det b_{\mu\nu}|}} .$$

Now, the hypothesis about the metricity of the background connection B is rather natural when metrics, which asymptote to a prescribed background metric $b_{\mu\nu}$, are considered. Further, the assumption in [38] that X is a Killing vector field of the background metric seems to be natural, and not overly restrictive, when X is thought of as representing time translations in the asymptotic region. Nevertheless, that last hypothesis is not adequate when one wishes to obtain simultaneously Hamiltonians for several vector fields X. Consider, for example, the problem of assigning a four-momentum to an asymptotically flat space-time — in that case four vector fields X, which asymptote to four linearly independent translations, are needed. The condition of invariance of the background metric $b_{\mu\nu}$ under a four-parameter family of flows generated by those vector fields puts then undesirable topological constraints on M. The situation is even worse when considering vector fields X, which asymptote to BMS supertranslations: in that case there are no background metrics, which are asymptotically flat and for which X is a Killing vector. This leads to the necessity of finding formulae in which neither the condition that $B^\mu_{\alpha\beta}$ is the Levi–Civita connection of some metric $b_{\alpha\beta}$, nor the condition that X is a Killing vector field of the background are imposed. It may be checked that the following generalization of equations (5.16)–(5.17) holds[2]:

$$p^\mu_{\alpha\beta}\pounds_X\mathfrak{g}^{\alpha\beta} - X^\mu L = \partial_\alpha\mathbb{W}^{\mu\alpha} - 2\mathfrak{g}^{\beta[\gamma}\delta^{\mu]}_\sigma(X^\sigma{}_{;\beta\gamma} - B^\sigma{}_{\beta\gamma\kappa}X^\kappa) , \tag{5.19}$$

$$\mathbb{W}^{\nu\lambda} = \left(2\mathfrak{g}^{\mu[\nu}p^{\lambda]}_{\mu\beta} - 2\delta^{[\nu}_\beta p^{\lambda]}_{\mu\sigma}\mathfrak{g}^{\mu\sigma} - \tfrac{2}{3}\mathfrak{g}^{\mu[\nu}\delta^{\lambda]}_\beta p^\sigma_{\mu\sigma}\right)X^\beta - 2\mathfrak{g}^{\alpha[\nu}\delta^{\lambda]}_\beta X^\beta{}_{;\alpha} \tag{5.20}$$

[2] The calculations of [38, Appendix] are actually done without any hypotheses on the vector field X; it is only at the end that it is assumed that X is a Killing vector field of the background metric. See also [90].

Here $B^\sigma{}_{\beta\gamma\kappa}$ is the curvature tensor of the connection $B^\sigma{}_{\beta\gamma}$.

It should be emphasized that equation (5.15) makes only sense for X's which are defined over M; in the general case, with X defined over $\mathscr{S}$ only, the Lie derivative $\pounds_X \mathfrak{g}^{\alpha\beta}$ isn't defined and one has to use the operator $\mathfrak{D}_X$ of equation (3.10) instead. It would be of interest to work out a general set of conditions under which the integrand of (5.15), with $\pounds_X$ replaced by $\mathfrak{D}_X$, can be written as the divergence of an anti-symmetric tensor density.

Consider the variational formula (3.43), and suppose for the sake of definiteness that $\mathscr{S}$ is a compact hypersurface with boundary. Recall that we have assumed that the motion of $\mathscr{S}$ arises from a vector field X defined on a neighbourhood of $\mathscr{S}$, and that $\mathfrak{D}_X p^\lambda{}_{\mu\nu}$ is the Lie derivative. Equation (3.43) takes the form

$$\begin{aligned} -\delta H = &\int_{\mathscr{S}} \left(\pounds_X p^\lambda{}_{\mu\nu}\delta\mathfrak{g}^{\mu\nu} - \pounds_X \mathfrak{g}^{\mu\nu}\delta p^\lambda{}_{\mu\nu}\right) dS_\lambda \\ &+ \int_{\partial\mathscr{S}} X^{[\mu}p^{\nu]}{}_{\alpha\beta}\delta\mathfrak{g}^{\alpha\beta}\, dS_{\mu\nu}\,, \end{aligned} \tag{5.21}$$

with $dS_{\mu\nu}$ defined in equation (3.44).

For the purposes of Sect. 6.2, let us note that the background connection can be completely eliminated from this formula. (On the other hand, we emphasize that the formula with the background metric is very convenient for most practical calculations here.) For this purpose we introduce the quantity

$$A^\alpha_{\mu\nu} := \Gamma^\alpha_{\mu\nu} - \delta^\alpha_{(\mu}\Gamma^\kappa_{\nu)\kappa}\,, \tag{5.22}$$

together with its counterpart for the background metric:

$$\mathring{A}^\alpha_{\mu\nu} := B^\alpha_{\mu\nu} - \delta^\alpha_{(\mu}B^\kappa_{\nu)\kappa}\,. \tag{5.23}$$

It follows that $p^\alpha_{\mu\nu} = \mathring{A}^\alpha_{\mu\nu} - A^\alpha_{\mu\nu}$. Now, we use the formula for the Lie derivative of a connection:

$$\pounds_X B^\lambda{}_{\mu\nu} = X^\lambda{}_{;\mu\nu} - X^\sigma B^\lambda{}_{\mu\nu\sigma}\,.$$

Consequently, we have:

$$\mathfrak{g}^{\mu\nu}\pounds_X \mathring{A}^\lambda{}_{\mu\nu} = 2\mathfrak{g}^{\beta[\gamma}\delta^{\lambda]}_\sigma\left(X^\sigma{}_{;\beta\gamma} - B^\sigma{}_{\beta\gamma\kappa}X^\kappa\right).$$

This, together with (5.19), shows that $H(X,\mathscr{S})$ given by (5.15) takes the form

$$H(X,\mathscr{S}) = H_{\mathrm{boundary}}(X,\mathscr{S}) + H_{\mathrm{volume}}(X,\mathscr{S})\,, \tag{5.24}$$

$$H_{\mathrm{boundary}}(X,\mathscr{S}) = \frac{1}{2}\int_{\partial\mathscr{S}} \mathbb{W}^{\alpha\mu}\, dS_{\alpha\mu}\,, \tag{5.25}$$

$$\begin{aligned} H_{\mathrm{volume}}(X,\mathscr{S}) &= -\int_{\mathscr{S}} \mathfrak{g}^{\mu\nu}\pounds_X \mathring{A}^\lambda{}_{\mu\nu}\, dS_\lambda \\ &= -\int_{\mathscr{S}} 2\mathfrak{g}^{\beta[\gamma}\delta^{\lambda]}_\sigma\left(X^\sigma{}_{;\beta\gamma} - B^\sigma{}_{\beta\gamma\kappa}X^\kappa\right) dS_\lambda\,. \end{aligned} \tag{5.26}$$

It follows that equation (5.21) can be rewritten as

$$-\delta H_{\text{boundary}}(X,\mathscr{S}) = \int_{\mathscr{S}} \left(\pounds_X \mathfrak{g}^{\mu\nu}\delta A^{\lambda}{}_{\mu\nu} - \pounds_X A^{\lambda}{}_{\mu\nu}\delta \mathfrak{g}^{\mu\nu}\right) dS_{\lambda} + \int_{\partial\mathscr{S}} X^{[\mu} p_{\alpha\beta}{}^{\nu]}\delta \mathfrak{g}^{\alpha\beta}\, dS_{\mu\nu} \,. \tag{5.27}$$

In this formula the Hamiltonian is a boundary integral, and the only background dependence in the right hand side of (5.27) is through the boundary terms.

5.2 Moving spacelike hypersurfaces

Before proceeding further, it is useful to recall the results of [94], which relate the above formulation of the theory to the usual ADM one. Suppose that $\mathscr{S}$ is a smooth spacelike hypersurface in (M,g). In a coordinate system in which $\mathscr{S}$ is a hypersurface of constant coordinate x^0, formula (5.14) reduces to:

$$\begin{aligned}\Omega_{\mathscr{S}}((\delta_1 p^{\lambda}_{\mu\nu},\delta_1\mathfrak{g}^{\alpha\beta}),(\delta_2 p^{\lambda}_{\mu\nu},\delta_2\mathfrak{g}^{\alpha\beta})) &= \int_{\mathscr{S}} (\delta_1 p^0_{\alpha\beta}\delta_2\mathfrak{g}^{\alpha\beta} - \delta_2 p^0_{\alpha\beta}\delta_1\mathfrak{g}^{\alpha\beta})\, d^3x \\ &= -\int_{\mathscr{S}} (\delta_1 A^0_{\alpha\beta}\delta_2\mathfrak{g}^{\alpha\beta} - \delta_2 A^0_{\alpha\beta}\delta_1\mathfrak{g}^{\alpha\beta})\, d^3x \,.\end{aligned} \tag{5.28}$$

We see that entire information about the momentum canonically conjugate to $\mathfrak{g}^{\alpha\beta}$ is encoded in 10 components $A^0_{\alpha\beta}$ of the connection Γ. It turns out that not all these components are independent of $\mathfrak{g}^{\alpha\beta}$ because equation (5.5) imposes some constraints on them. More precisely, equations (5.5) for $\partial_k\mathfrak{g}^{00}$ and $\partial_k\mathfrak{g}^{k0}$ are equivalent to the following four constraints (see [94]):

$$A^0_{00} = \frac{1}{\mathfrak{g}^{00}}\left(\partial_k\mathfrak{g}^{0k} + A^0_{kl}\mathfrak{g}^{kl}\right) , \tag{5.29}$$

$$A^0_{0k} = -\frac{1}{2\mathfrak{g}^{00}}\left(\partial_k\mathfrak{g}^{00} + 2A^0_{kl}\mathfrak{g}^{0l}\right) . \tag{5.30}$$

Inserting these constraints into the volume integral (5.27) and into the form (5.28), we obtain the reduced version of the theory. It is convenient to introduce the following notation:

$$P^{kl} := \sqrt{\det g_{mn}}\,\left({}^3g^{ij}K_{ij}\,{}^3g^{kl} - K^{kl}\right) , \tag{5.31}$$

where

$$K_{kl} := -\frac{1}{\sqrt{|g^{00}|}}\Gamma^0_{kl} = -\frac{1}{\sqrt{|g^{00}|}}A^0_{kl} , \tag{5.32}$$

with ${}^3g^{kl}$ — the three-dimensional inverse of the induced metric g_{kl} on $\mathcal{S}$; the indices on K^{kl} have been raised using ${}^3g^{kl}$. Suppose, in addition, that $\mathcal{S}$ has compact closure and smooth compact boundary $\partial\mathcal{S}$. Choose the coordinate x^3 in such a way that $\mathcal{S} \cap \mathcal{O} = \{x^0 = 0\}$, and $\partial\mathcal{S} \cap \mathcal{O} = \{x^0 = 0, x^3 = 1\}$. In terms of these quantities the "symplectic" form (5.28) reduces to:

$$\begin{aligned}
&\Omega_{\mathcal{S}}((\delta_1 p^\lambda_{\mu\nu}, \delta_1 \mathfrak{g}^{\alpha\beta}), (\delta_2 p^\lambda_{\mu\nu}, \delta_2 \mathfrak{g}^{\alpha\beta})) \\
&= \frac{1}{16\pi} \int_{\mathcal{S}} \left(\delta_1 g_{kl} \delta_2 P^{kl} - \delta_2 g_{kl} \delta_1 P^{kl}\right) d^3x \\
&\quad + \frac{1}{16\pi} \int_{\partial\mathcal{S}} \left(\delta_1 N^3 \delta_2 \frac{\sqrt{\det g_{kl}}}{N} - \delta_2 N^3 \delta_1 \frac{\sqrt{\det g_{kl}}}{N}\right) d^2x \,, \qquad (5.33)
\end{aligned}$$

where

$$N = 1/\sqrt{-g^{00}} \,, \quad N_k = g_{0k} \,, \quad N^3 = (g^{3k} - \frac{g^{03}g^{0k}}{g^{00}}) N_k \,.$$

The volume integral here gives the standard ADM symplectic form. It was noticed in [92, 93] that the boundary term gives the necessary correction, which makes the form invariant with respect to all diffeomorphisms, which do not move the boundary $\partial\mathcal{S}$.

The tensor fields (g_{kl}, P^{kl}) on $\mathcal{S}$ are still not arbitrary, as they fulfill the four Gauss–Codazzi constraint equations:

$$P_i{}^l{}_{|l} = 8\pi \sqrt{\det g_{mn}}\, T_{i\mu} n^\mu \,, \qquad (5.34)$$

$$(\det g_{mn}) \mathcal{R} - P^{kl} P_{kl} + \tfrac{1}{2}(P^{kl} g_{kl})^2 = 16\pi (\det g_{mn}) T_{\mu\nu} n^\mu n^\nu \,. \qquad (5.35)$$

Here $T_{\mu\nu}$ is an energy-momentum tensor of matter fields (if any), by $\mathcal{R}$ we denote the (three-dimensional) scalar curvature of g_{kl}, $n = n^\mu \partial_\mu$ is a future timelike four-vector normal to the hypersurface $\mathcal{S}$, "$|$" denotes the three-dimensional covariant derivative induced by g_{kl}, indices have been raised with ${}^3g^{ij}$.

Let ψ and Σ be as in Chap. 3, we define a τ-dependent one-parameter family of symmetric tensor field $\gamma(\tau, \cdot)$ on Σ by[3]

$$\gamma(\tau, \cdot) = \gamma_{ij}(\tau, \cdot) dx^i dx^j = (\psi_\tau)^* g \,. \qquad (5.36)$$

We shall make the hypothesis that for each τ the image $\psi_\tau(\Sigma)$ is a spacelike hypersurface in M — this is equivalent to the requirement that for each τ the tensor field $\gamma(\tau, \cdot)$ defines a Riemannian metric on Σ. Next, let $\nu(\tau, \cdot)$ be a τ-dependent one-parameter family of functions on Σ defined by

$$\nu(\tau, \cdot) = g(\psi_* \frac{\partial}{\partial \tau}, n_\tau) \circ \psi_t(\cdot) \,. \qquad (5.37)$$

[3] As in Chap. 3 we use different symbols to differentiate between a field on $\mathcal{S}$ — in the current case the metric g_{ij} — and its pull-back γ_{ij} to the model space Σ.

Here n_τ is a future timelike four-vector normal to the hypersurface $\psi_\tau(\Sigma)$. (The function $\nu \circ \psi$ is often called the lapse function in the literature, and is often denoted by N.) We note that the vector field

$$\psi_* \frac{\partial}{\partial \tau} + \nu \circ \psi \, n_\tau$$

is orthogonal to n_τ, hence tangent to $\psi_\tau(\Sigma)$, thus there exists a τ-dependent one-parameter family of vector fields $\beta(\tau, \cdot)$ on Σ such that

$$\psi_* \frac{\partial}{\partial \tau} + \nu \circ \psi \, n_\tau = (\psi_\tau)_* \beta(\tau, \cdot) \,. \tag{5.38}$$

(The vector field $(\psi_\tau)_*\beta(\tau,\cdot)$ is often called the shift in the literature, and is often denoted by $N^i\partial_i$.) A standard calculation shows that in local coordinates on $I \times \Sigma$ we have

$$\psi^* g = -\nu^2 d\tau^2 + \gamma_{ij}(dx^i + \beta^i d\tau)(dx^j + \beta^j d\tau) \,. \tag{5.39}$$

We emphasize that $\psi^* g$ is in general *not* a Lorentzian metric on $I \times \Sigma$ — at those points at which ψ_* degenerates the tensor field $\psi^* g$ will obviously degenerate as well. Now, those points are precisely the ones at which ν has zeros. However, the vanishing of ν has nothing to do with the question whether or not the tensor field g itself is degenerate (and hence is a Lorentzian metric on M). We mention this fact because in most expositions of the ADM framework it is assumed at the outset that the coordinates (τ, x^i) of equation (5.39), which are coordinates on $I \times \Sigma$, also form a coordinate system on $\psi(I \times \Sigma)$. This creates unnecessary conceptual difficulties when dealing with zeros of $\nu \circ \psi$, which are avoided altogether when the framework is understood in the way proposed here.

Let us for completeness give the set of equations satisfied by γ_{kl} and π^{kl} [100, p. 525][4]:

$$\dot\gamma_{kl} = \frac{2\nu}{\sqrt{\det(\gamma_{ij})}} \left(\pi_{kl} - \frac{1}{2}\gamma_{kl}\pi\right) + \beta_{k|l} + \beta_{l|k} \,, \tag{5.40}$$

where $\pi := \pi^{kl}\gamma_{kl}$; a dot denotes a τ-derivative. Further

$$\begin{aligned}
\dot\pi^{kl} &= -\nu\sqrt{\det(\gamma_{ij})}\left(\mathscr{R}^{kl} - \frac{1}{2}\gamma^{kl}\mathscr{R}\right) - \frac{2\nu}{\sqrt{\det(\gamma_{ij})}}\left(\pi^{km}\pi_m{}^l - \frac{1}{2}\pi\pi^{kl}\right) \\
&\quad + \left(\pi^{kl}\beta^m\right)_{|m} + \frac{\nu}{2\sqrt{\det(\gamma_{ij})}}\gamma^{kl}\left(\pi^{mn}\pi_{mn} - \frac{1}{2}\pi^2\right) - \beta^k{}_{|m}\pi^{ml} - \beta^l{}_{|m}\pi^{mk} \\
&\quad + \sqrt{\det(\gamma_{ij})}\left(\nu^{|kl} - \gamma^{kl}\nu^{|m}{}_{|m}\right) + 8\pi\nu\sqrt{\det(\gamma_{ij})}T_{mn}\gamma^{km}\gamma^{ln} \,,
\end{aligned} \tag{5.41}$$

[4] It is an instructive exercise to derive those equations via calculations in which ν^{-1} never appears, so that one does not need to assume that ν has no zeros.

where $\mathscr{R}_{ij}$ is the Ricci tensor of γ_{ij}. The right hand sides of those equations are smooth independently of whether or not ν has zeros. It should be recognized that when one tries to *construct* a space-time using PDE methods it is convenient (and often unavoidable) to use a ψ in which the corresponding ν function has no zeros. This, however, has nothing to do with the problems addressed here, because one can use ψ's, which form a coordinate system to construct (M, g) (using PDE theory) and, once (M, g) has been constructed, use ψ's which do not necessarily lead to coordinate systems on M to obtain Hamiltonian dynamical systems on appropriate phase spaces.

The Hamiltonian form in (5.27) may also be reduced in a way analogous to the reduction (5.33) of the symplectic form (5.28) (see [94]). For this purpose we have only to replace δ_1 by the Lie derivative with respect to X. The latter reduces to the time derivative on the parameter space-time $I \times \Sigma$, and we denote it as usual by a "dot". In this way we obtain:

$$\int_{\psi_0(\Sigma)} (\pounds_X \mathfrak{g}^{\alpha\beta} \delta A^\mu_{\alpha\beta} - \pounds_X A^\mu_{\alpha\beta} \delta \mathfrak{g}^{\alpha\beta})\, dS_\mu = \\ \frac{1}{16\pi} \int_\Sigma \left(\dot{\pi}^{kl} \delta\gamma_{kl} - \dot{\gamma}_{kl} \delta\pi^{kl}\right) d^3x + \frac{1}{16\pi} \int_{\partial\Sigma} (\dot{\mu}\delta\sigma - \dot{\sigma}\delta\mu) d^2x \,, \quad (5.42)$$

where $\sigma := \beta^3$ (the x^3 component of the shift vector) and $\mu := \sqrt{\det \gamma_{kl}}/\nu$. This equation, inserted into (5.27), gives

$$-\delta H_{\text{boundary}}(X, \mathscr{S}) = \frac{1}{16\pi} \int_\Sigma \left(\dot{\pi}^{kl} \delta\gamma_{kl} - \dot{\gamma}_{kl} \delta\pi^{kl}\right) d^3x \\ + \frac{1}{16\pi} \int_{\partial\Sigma} (\dot{\mu}\delta\sigma - \dot{\sigma}\delta\mu) d^2x + \int_{\partial\psi_0(\Sigma)} (X^0 p^3_{\alpha\beta} \delta\mathfrak{g}^{\alpha\beta} - X^3 p^0_{\alpha\beta} \delta\mathfrak{g}^{\alpha\beta})\, d^2x \quad (5.43)$$

We conclude that $H_{\text{boundary}}(X, \mathscr{S})$ is also a Hamiltonian for the ADM symplectic form, in situations in which the boundary integrals on the right hand side vanish. We study this question in detail in Sect. 5.4 below when asymptotic flatness at spatial infinity is assumed; in that case $\mathscr{S}$ is not compact but appropriate limits can be easily justified. If the background connection is chosen to be the Levi–Civita connection of a metric which has constant coefficients in an asymptotically Minkowskian coordinate system, and if X is taken to be asymptotic (at an appropriate rate) to a time translation, then (5.15) coincides with the ADM energy [38].

In a general situation, when Σ is a compact manifold with boundary, a further reduction of the last term in the above formula was performed in [94]. The reduction is based on the metricity conditions for $A^3_{3\alpha}$, analogous to (5.29) and (5.30), where the coordinate x^3 is constant on $\partial\Sigma$. This way the ADM momentum of the three-manifold $\psi_0(I \times \partial\Sigma)$ arises, and the boundary (correction) term in the symplectic form is rewritten in terms of a (hyperbolic) angle between this manifold and the Cauchy manifold $\psi_0(\Sigma)$. We are, however, not going to need this formula in the present work.

5.3 Cosmological space-times

The simplest illustration of the framework of Chap. 3 occurs, for gravitating systems, in the context of cosmological space-times — by definition, these are space-times with compact boundaryless Cauchy hypersurfaces. The discussion is much simpler here due to the fact that for smooth fields no convergence problems arise; further, all boundary integrals vanish. Let, then, Σ be a three dimensional compact manifold without boundary, we define $\mathscr{P}_{\mathrm{cosm}}$ to be the set of all pairs (γ, π) where, as in the previous section, γ is a Riemannian metric on Σ, and π is a symmetric tensor density on Σ. Further we require that (γ, π) satisfy the vacuum constraint equations — *i.e.*, equations (5.34)-(5.35), with g_{ij} and P^{ij} there replaced by γ_{ij} and π^{ij}, and with $T_{\mu\nu} = 0$. Pairs (γ, π) satisfying those conditions will be called *vacuum initial data sets.* Standard theory (*cf.*, *e.g.*, [33, 35] or [119, Vol. III]) associates to each data set in $\mathscr{P}_{\mathrm{cosm}}$ a vacuum, maximal globally hyperbolic space-time $(M_{(\gamma,\pi)}, g)$, unique up to diffeomorphism, together with an embedding

$$i : \Sigma \to M_{(\gamma,\pi)} ,$$

such that the pull-back i^*g of g to Σ equals γ, and such that $\pi = i^*P$, with P associated to the extrinsic curvature tensor of $i(\Sigma)$ as in Equation (5.31). (This holds actually for any manifold Σ, the requirement of compactness is not needed here.) We should write $g_{(\gamma,\pi)}$ for g, but we shall stick to g to avoid an overburdening of notation. Now, the Hamiltonian formalism developed in Chap. 3 makes use of maps ψ from $I \times \Sigma$ into a fixed manifold M, *cf.* Equation (3.5), rather than the family of manifolds $M_{(\gamma,\pi)}$ obtained so far. The usual way of obtaining a single manifold M is by choosing a diffeomorphism

$$\psi_{(\gamma,\pi)} : \mathscr{O}_{(\gamma,\pi)} \longrightarrow (-t_-, t_+) \times \Sigma , \tag{5.44}$$

where $\mathscr{O}_{(\gamma,\pi)}$ is a neighbourhood of $i(\Sigma)$ in $M_{(\gamma,\pi)}$, for some $t_- < 0 < t_+$ depending in general upon the initial data. When considering time translations the map ψ of (3.5) is then simply the identity

$$\psi = \mathrm{id}_{I\times\Sigma} .$$

The key role here is played of course by the families of maps $\psi_{(\gamma,\pi)}$, which are determined by so-called "gauge-fixing" procedures. There is considerable freedom in the choice of the $\psi_{(\gamma,\pi)}$'s, and each such choice leads to a different identification between flows of vectors X on M and corresponding motions in the space-times $M_{(\gamma,\pi)}$. We will discuss here one such possibility, the "wave map reduction" of the vacuum Einstein equations. This reduction uses a background metric b, which makes it well adapted to the background metric reduction of the Lagrangian presented in Sect. 5.1. Let, then, b be any fixed Lorentzian metric on $\mathbb{R} \times \Sigma$; a possible choice is

$$b = -dt^2 + \mathring{\gamma} , \tag{5.45}$$

where $\mathring{\gamma}$ is some fixed Riemannian metric on Σ. We shall, however, not assume that b is of the form (5.45), because there are situations in which other choices may be physically justified: For example, in perturbation theory it might be convenient to take b to be one of the Robertson-Walker metrics. The map $\psi_{(\gamma,\pi)}$ of Equation (5.44) will be chosen to satisfy the wave map equation, when $M_{(\gamma,\pi)}$ is equipped with the vacuum metric g, and $\mathbb{R}\times\Sigma$ is equipped with the metric b; by definition, in local coordinates, we have

$$\square_g \psi^\mu = {}^{\psi}B^\mu_{\nu\rho}\, g^{\alpha\beta}\frac{\partial\psi^\nu}{\partial x^\alpha}\frac{\partial\psi^\rho}{\partial x^\beta} \;. \tag{5.46}$$

Here the ${}^{\psi}B^\mu_{\nu\rho}$'s are the Christoffel symbols $B^\mu_{\nu\rho}$ of the metric b composed with ψ. We shall also impose the initial conditions

$$\begin{aligned} \psi^0(t=0,\cdot) &= 0\;, \quad & n^\mu\frac{\partial\psi^0}{\partial x^\mu}(t=0,\cdot) &= 1\;, \\ \psi^i(t=0,\cdot) &= \mathrm{id}_\Sigma\;, \quad & n^\mu\frac{\partial\psi^i}{\partial x^\mu}(t=0,\cdot) &= 0\;. \end{aligned} \tag{5.47}$$

Here n^μ is the field of unit, future directed normals to $i(\Sigma)$. It follows from standard results on hyperbolic PDE's (*cf.*, *e.g.*, [119, Vol. III]) that for smooth initial data there exists a neighbourhood $\mathscr{O}_{(\gamma,\pi)}$ of $i(\Sigma)$ in $M_{(\gamma,\pi)}$ on which $\psi_{(\gamma,\pi)}$ is a diffeomorphism as in (5.44). Transporting the metric g to $(t_-,t_+)\times\Sigma$ using $\psi_{(\gamma,\pi)}$ one obtains on a subset of M the desired vacuum metric — still denoted by g, solution of the Cauchy problem with initial data (γ,π).

Now, any map

$$\psi : I\times\Sigma \longrightarrow M := \mathbb{R}\times\Sigma$$

with the property that the images $\psi_t(\Sigma)$ are spacelike hypersurfaces can be used to define a local dynamical system on $\mathscr{P}_{\mathrm{cosm}}$ by setting

$$T_t(\gamma,\pi) := (\gamma(t) := \psi_t^* g, \pi(t)) \;.$$

Here, as in Chap. 3, $\psi_t(x) := \psi(t,x)$, while $\pi(t)$ is related to the extrinsic curvature $K(t)$ of the hypersurfaces $\psi_t(\Sigma)$ as in Equation (5.31) with g_{ij} replaced by γ_{ij}. When ψ arises from the flow of a vector field X on M,

$$\frac{d\psi_t}{dt} = X\circ\psi_t \;,$$

we then have the variational formula (5.21), with the Hamiltonian given by Equations (5.24)-(5.26); as already pointed out, all the boundary integrals there vanish. The volume integral (5.26) vanishes as well when X is a Killing vector field of the background metric. Thus, the Hamiltonians are trivial for background Killing vectors, even though the dynamics is not. We emphasize that the Hamiltonians *do not* vanish in general for X's which are not b-Killing vector fields; for example, on a Robertson-Walker background the Hamiltonian associated with time translations will be non-trivial in the context of Equation (5.21).

An alternative variational formula is (5.27), with vanishing Hamiltonian regardless of whether or not the vector field X is a Killing vector field for the background metric b . Similarly, the usual ADM variational formula (5.43) always leads to vanishing Hamiltonians in space-times with compact boundaryless Cauchy surfaces, whatever the vector field X.

5.4 Space-times asymptotically flat in spacelike directions

The next example we consider is that of space-times which are asymptotically flat in spacelike directions. The problem is somewhat more complicated, as compared to the cosmological one, because of the non-compactness of the spacelike hypersurfaces under consideration; this leads to convergence questions, and to the occurrence of boundary terms at infinity. On the other hand it is considerably simpler than the problem we are ultimately interested in — energy in the radiation regime — because there is no outflow, or inflow, of energy in the asymptotic region in this case.

In the setting here the manifold Σ is assumed to be the union of a compact manifold Σ_{int} together with an asymptotic region Σ_{ext} diffeomorphic to $\mathbb{R}^3 \setminus B(0,R)$, where $B(0,R)$ denotes a coordinate ball of radius R centered at the origin. Thus we assume that there is only one asymptotically flat end Σ_{ext}; the generalisation to a finite number of ends is straightforward and is left to the reader — one obtains then a mass, momentum, *etc.*, for each asymptotically flat region.

On $\mathbb{R} \times \Sigma$ we choose any metric b such that

$$b = -dt^2 + dx^2 + dy^2 + dz^2 \quad \text{on } \mathbb{R} \times \mathbb{R}^3 \setminus B(0,2R) \,, \tag{5.48}$$

in the coordinate system in which (5.49)-(5.50) hold. The coordinates $(x^\mu) \equiv (t,x,y,z)$ of (5.48) will be referred to as *quasi-Minkowskian coordinates.* The standard way [16, 40, 101] of defining asymptotic flatness associated to the background b is to require that in the quasi-Minkowskian coordinates on $\mathbb{R} \times \Sigma_{\text{ext}}$ the physical metrics g satisfies

$$|g_{\mu\nu} - \eta_{\mu\nu}| = o(r^{-1/2}) \,, \quad |\partial_\sigma g_{\mu\nu}| = o(r^{-3/2}) \,, \tag{5.49}$$

$$\partial_\sigma g_{\mu\nu} \in L^2(\Sigma_{\text{ext}}) \,. \tag{5.50}$$

We note that those conditions require a map $\psi_{(\gamma,\pi)}$ as in (5.44), to transport the metric from the physical space-time to $\mathbb{R} \times \Sigma$, we will return to this question shortly. For example, Equations (5.49)-(5.50) will hold if we have

$$|g_{\mu\nu} - \eta_{\mu\nu}| = O(r^{-\alpha}) \,, \quad |\partial_\sigma g_{\mu\nu}| = O(r^{-\alpha-1}) \,, \quad \text{for some } \alpha > 1/2 \,, \tag{5.51}$$

and these equations are satisfied in Schwarzschild space-time with $\alpha = 1$. Under reasonably mild supplementary conditions (involving weighted-Sobolev conditions on derivatives of order two of the metric), the positive energy theorem [20, 81, 82, 85, 112, 127] implies that the decay rate $\alpha = 1$ cannot be improved for vacuum space-times except for initial data sets corresponding to the Minkowski space-time. However, under (5.51) the integrand of (5.14) behaves as $r^{-1-2\alpha}$, which in space-dimension 3 leads to a divergent integral for $\alpha \leq 1$. This shows that a more careful analysis of this integrand is necessary. There are various ways of proceeding, in the approach presented here we shall exploit the observation, originally due to Lichnerowicz, that the scalar constraint equation (5.35) becomes an elliptic equation for a conformal factor once a conformal class of γ_{ij} has been prescribed. We thus write

$$\gamma_{ij} = e^{\phi}\widehat{\gamma}_{ij} , \tag{5.52}$$

so that the volume integrand in (5.33) takes the form

$$\begin{aligned} &\delta_1\gamma_{kl}\, \delta_2\pi^{kl} - \delta_1 \leftrightarrow \delta_2 \\ &\quad = \delta_1\phi\, \delta_2(\gamma_{kl}\pi^{kl}) - e^{\phi}\left[\pi^{kl}\delta_1\phi\, \delta_2\widehat{\gamma}_{kl} - \delta_1\widehat{\gamma}_{kl}\, \delta_2\pi^{kl}\right] - \delta_1 \leftrightarrow \delta_2 \, . \end{aligned} \tag{5.53}$$

The point here is that while the fall-of of ϕ cannot be better than $1/r$ in general, that of $\widehat{\gamma}_{ij}$ may be chosen at will. One is then left with only one problematic term in (5.53), namely $\delta_1\phi\, \delta_2(\gamma_{kl}\pi^{kl}) - \delta_1 \leftrightarrow \delta_2$. We will control this term by imposing stronger restrictions on $\gamma_{kl}\pi^{kl}$ than on π^{kl} or, equivalently, by imposing further conditions on the trace of the extrinsic curvature tensor. Choose any vacuum initial data set $(\mathring{\gamma}_{ij}, \mathring{\pi}^{kl})$ arising from a space-time satisfying the conditions (5.49)-(5.50), and let $\mathring{\pi}$ be the trace of the ADM momentum tensor $\mathring{\pi}^{kl}$:

$$\mathring{\pi} = \mathring{\gamma}_{kl}\mathring{\pi}^{kl} \, . \tag{5.54}$$

(One easily checks, using Equations (5.40) and (5.49), that $\mathring{\pi} = o(r^{-3/2})$.) In this section we will explicitly put indices on all tensorial objects so that $\mathring{\pi}$ is *not* the same object as the tensor $\mathring{\pi}^{kl}$; we hope that the reader will not get confused by the fact that those indices are not explicitly displayed in other sections. We will start by defining a phase space $\mathscr{P}_{\mathring{\pi}}$ by imposing a series of requirements, the desired phase space $\mathscr{P}_{\mathrm{ADM}}$ will then be defined by Equation (5.60) below. First, we shall assume that all initial data (γ_{ij}, π^{kl}) in $\mathscr{P}_{\mathring{\pi}}$ satisfy

$$\phi = O(r^{-1}) \, , \quad \frac{\gamma_{kl}\pi^{kl} - \mathring{\pi}}{r} \in L^1(\Sigma_{\mathrm{ext}}) \, . \tag{5.55}$$

This will hold if, *e.g.*,

$$\phi = O(r^{-1}) \, , \quad \gamma_{kl}\pi^{kl} - \mathring{\pi} = O(r^{-\gamma}) \, , \quad \text{for some } \gamma > 2 \, . \tag{5.56}$$

In view of (5.53), the convergence of the integral in (5.14) or — equivalently — of the ADM form (5.33), will hold (by the Bunyakovsky-Cauchy-Schwartz inequality in $L^2(\Sigma_{\mathrm{ext}})$) if we further assume that

$$\widehat{\gamma}_{ij} - \delta_{ij} \, , \pi^{kl} \in L^2(\Sigma_{\mathrm{ext}}) \, . \tag{5.57}$$

This will be satisfied, *e.g.*, for fields such that

$$\widehat{\gamma}_{ij} - \delta_{ij} = O(r^{-\beta}) \, , \quad \pi^{ij} = O(r^{-\beta}) \, , \qquad \text{for some } \beta > 3/2 \, . \tag{5.58}$$

We shall finally impose on $\mathscr{P}_{\mathring{\pi}}$ some conditions which will arise — in an implicit way — from the requirement that $\mathscr{P}_{\mathring{\pi}}$ be preserved under evolution by the vacuum Einstein equations. This is related to the question of the "choice of gauge", that is, of the map in Equation (5.44). We emphasize that even under the restriction (5.49), (5.50) and (5.55) there is considerable arbitrariness left. Further, the Hamiltonians "associated with a vector field $\partial/\partial t$" *do* depend upon the choice made, because the field $\partial/\partial t$ in one such choice will in general *not* coincide with the vector $\partial/\partial t$ arising from another choice of the maps (5.44). Nevertheless, it can be shown [16, 40, 41] that the numerical value of the Hamiltonians (5.61)-(5.62) below, associated with space-time translations, is independent of this choice as long as the asymptotic conditions (5.51) hold.

Let, as in the previous section, $(M_{(\gamma,\pi)}, g)$ be the maximal globally hyperbolic development of any initial data $(\gamma_{ij}, \pi^{kl}) \in \mathscr{P}_{\mathring{\pi}}$, and let τ be any time function defined on a globally hyperbolic neighbourhood in $M_{(\gamma,\pi)}$ of $i(\Sigma)$ (by definition, a *time function* τ is a differentiable function with the property that its gradient $\nabla\tau$ is timelike). We also assume that $i(\Sigma)$ is a Cauchy surface for this neighbourhood. There is then a natural diffeomorphism of this neighbourhood and a subset of $\mathbb{R} \times \Sigma$ obtained by assigning to p the couple (t, x), where $t = \tau(p)$, while x is defined as the intersection with $i(\Sigma)$ of the integral curve of ∇t passing through p. Let $(\gamma_{ij}(t), \pi^{kl}(t))$ be the Cauchy data induced by g on the level set $\{\tau(p) = t\}$ of τ. The requirement that the dynamics preserves $\mathscr{P}_{\mathring{\pi}}$ imposes the restriction that (5.55) holds with (γ_{ij}, π^{kl}) replaced by $(\gamma_{ij}(t), \pi^{kl}(t))$; this will be guaranteed if we choose the function τ so that we have

$$\gamma_{ij}(t,x)\pi^{ij}(t,x) = \gamma_{ij}(0,x)\pi^{ij}(0,x) \, . \tag{5.59}$$

This gives then one possible proposal[5] how to choose a desired map in (5.44).

Several comments are in order here. First, (5.59) is, essentially, the *prescribed mean curvature equation.* Its linearization leads to an elliptic, second order equation, which has no kernel in vacuum. This means, in our asymptotically flat context, that the solutions of (5.59) with the asymptotic requirement that τ tends to a constant when $r(x)$ tends to infinity do lead to a foliation of a neighbourhood $\mathscr{O}_{(\gamma,\pi)}$ of $i(\Sigma)$, whenever they exist. Next, under some mild supplementary hypotheses on the initial data, involving weighted Hölder or

[5] One could use the wave-map condition (5.46) here, if desired; this requires checking that the associated evolution preserves the conditions imposed, which might lead to the need of imposing some further conditions on $\mathscr{P}_{\mathring{\pi}}$ to obtain a consistent scheme. We have not attempted to analyse this question.

Sobolev conditions on a finite number of derivatives of the metric, one can use the boost theorem of Christodoulou and O'Murchadha [37] together with the implicit function theorem to obtain existence of the required neighbourhood $\mathscr{O}_{(\gamma,\pi)}$ (compare [15, 18]), so that the coordinates (t,x) cover a set of the form $(t_-,t_+)\times\Sigma$, for some $t_-<0<t_+$, and so that the fall-off conditions (5.49)-(5.50) and (5.57) hold on the level sets of t. Rather than writing down the list of supplementary hypotheses needed for all those properties to hold, as our final restriction on $\mathscr{P}_{\mathring{\pi}}$ we shall require instead that *the fields in* $\mathscr{P}_{\mathring{\pi}}$ *be such that conditions (5.49), (5.50), and (5.57) are preserved by the evolution implicitly defined by Equation (5.59).*

It should be pointed out that the usual method of constructing solutions of the vacuum constraint equations imposes $\gamma_{ij}K^{ij}=0$ at the outset [35] (*cf.*, however, [34, 79]), so that (5.59) does not seem to be overly restrictive from this point of view. Nevertheless it is unsatisfactory that each space $\mathscr{P}_{\mathring{\pi}}$ contains only a subset of all the initial data arising from space-times satisfying the asymptotic conditions (5.49), (5.50) and (5.57). This is cured by setting

$$\mathscr{P}_{\mathrm{ADM}}:=\cup_{\mathring{\pi}}\mathscr{P}_{\mathring{\pi}}\ , \tag{5.60}$$

where the union is taken over all smooth functions on Σ satisfying $\mathring{\pi}=o(r^{-3/2})$, together with perhaps some finite number of weighted conditions of the derivatives of $\mathring{\pi}$ related to the structure of the spaces $\mathscr{P}_{\mathring{\pi}}$. It is natural to view $\mathscr{P}_{\mathrm{ADM}}$ as a manifold, with the $\mathscr{P}_{\mathring{\pi}}$'s providing coordinate patches, in the following loose sense: It follows from what has been said about the mean curvature equation that for each $(\gamma_{ij},\pi^{kl})\in\mathscr{P}_{\mathring{\pi}}$ there is an open neighbourhood $\mathscr{O}$ of the function $\mathring{\pi}$ (in an appropriate weighted Sobolev or weighted Hölder topology) such that for any function $\pi\in\mathscr{O}$ there exists an asymptotically flat hypersurface $\mathscr{S}'$ in $M_{(\gamma,\pi)}$ with the property that the metric γ'_{ij} and the ADM momentum π'^{kl} induced by the space-time metric g on $\mathscr{S}'$ satisfy

$$\gamma'_{ij}\pi'^{ij}=\pi\ .$$

In other words, the data induced on $\mathscr{S}'$ are in $\mathscr{P}_{\pi}$. Thus, dynamics by the vacuum Einstein equations allows one to pass from one space $\mathscr{P}_{\mathring{\pi}}$ to another, within open sets of functions $\mathring{\pi}$.

We have already established convergence of the volume integral in Equation (5.33), the vanishing of the boundary integral there immediately follows from (5.49). Let X be a translation in the asymptotic region; that is, $X=a^\mu\partial_\mu$ has constant coefficients a^μ in the coordinates of (5.48). One then easily checks that under (5.49)-(5.50) the boundary integrals in Equation (5.43) vanish, and the volume integral converges. General arguments can be given [126] that this implies the convergence of $H_{\mathrm{boundary}}(X,\mathscr{S})$; an alternative is to verify directly from the defining formula (5.15) that the volume integral there converges in view of (5.49)-(5.50). This establishes the Hamiltonian character, in the sense of (5.43), with Hamiltonian $H_{\mathrm{boundary}}(X,\mathscr{S})$, of space-time translations for fields in $\mathscr{P}_{\mathrm{ADM}}$. Similarly the variational formula (5.27) holds with a convergent volume integral and vanishing boundary integral. Finally, if $a^\mu\partial_\mu$ is a Killing vector of the background metric b, then $H_{\mathrm{boundary}}(X,\mathscr{S})$ is again a Hamiltonian in the sense of Equation (5.21) with no boundary term there; otherwise the (convergent) volume contribution (5.26) has to be added to obtain a Hamiltonian dynamics in the sense of (5.21), still with no boundary term.

For any set of constants a^μ we define the Arnowitt-Deser-Misner four-momentum p^μ as

$$p_\mu a^\mu := -H_{\mathrm{boundary}}(a^\mu\partial_\mu,\mathscr{S})\ ,$$

where $\mathscr{S} = i(\Sigma)$; the minus sign here is due to our signature $(-+++)$. A somewhat lengthy calculation leads to the standard ADM [7,8] form of energy-momentum of vacuum gravitational fields satisfying the fall-off conditions (5.49)-(5.50); in terms of fields on $i(\Sigma)\subset M$ we have:

$$m \equiv p^0 := \frac{1}{16\pi}\int_{S^2}(g_{ij,j}-g_{jj,i})d\sigma_i\ , \tag{5.61}$$

$$p^i := \frac{1}{8\pi}\int_{S^2}P^{ij}d\sigma_j\ . \tag{5.62}$$

The reader is referred to [21, 107] for an alternative Hamiltonian treatment of asymptotically flat space-times, and to [3, 17] for an analysis of some related analytical issued.

To close this section, we mention that the requirement of a finite and well defined angular momentum imposes some further restrictions on the asymptotic behaviour of the fields under consideration, see [21, 36, 39, 84].

5.5 Space-times with anti-de Sitter asymptotic behaviour

As another illustration of the geometric Hamiltonian techniques, we consider solutions of vacuum Einstein equations with a *strictly negative* cosmological constant Λ. We will restrict our attention to metrics g that asymptote to background metrics b of the form

$$b = -\left(\frac{r^2}{\ell^2}+k\right)dt^2 + \left(\frac{r^2}{\ell^2}+k\right)^{-1}dr^2 + r^2h\ ,$$
$$h = h_{AB}(v^C)dv^Adv^B\ ,\quad k=\mathrm{const}\ , \tag{5.63}$$

where[6] h is a Riemannian metric on $\check{M}$, and ℓ is a strictly positive constant related to the cosmological constant Λ by the formula

$$2\Lambda = -n(n-1)/\ell^2 \,. \tag{5.64}$$

The manifold $\check{M}$ will be assumed to be compact (without boundary) throughout this section. For example, if h is the standard round metric on S^2 and $k = 1$, then b is the anti-de Sitter metric. Examples of non-trivial metrics g asymptotic to the anti-de Sitter metric are the Kottler [96] ("Schwarzschild – anti-de Sitter") metrics

$$g = -\left(k - \frac{2m}{r} + \frac{r^2}{\ell^2}\right) dt^2 + \left(k - \frac{2m}{r} + \frac{r^2}{\ell^2}\right)^{-1} dr^2 + r^2 h \,, \tag{5.65}$$

where k is again equal to one, m is a constant, and h is a unit round metric on S^2. Another family of examples is given by the "generalized Kottler metrics" [28, 124], again of the form (5.65), except that h is now a Riemannian metric on a higher genus two dimensional manifold $\check{M}$. These metrics are solutions of the vacuum Einstein equations with a cosmological constant given by formula (5.64) if and only if h has constant Gauss curvature equal to k. We mention that the Hamiltonian mass defined by Equation (5.87) below coincides with the mass parameter m in (5.65) when h is the round unit sphere, is proportional to the parameter m in (5.65) when h is a metric of constant Gauss curvature k on some higher genus manifold $\check{M}$, and is proportional to the parameter η which occurs in the $(n+1)$-dimensional generalizations of the Kottler metrics [78]

$$g = -\left(1 - \frac{2\eta}{r^{n-2}} + \frac{r^2}{\ell^2}\right) dt^2 + \left(1 - \frac{2\eta}{r^{n-2}} + \frac{r^2}{\ell^2}\right)^{-1} dr^2 + r^2 h \,, \tag{5.66}$$

with h — a round metric of scalar curvature $(n-1)(n-2)$ on a $(n-1)$-dimensional sphere (*cf.*, *e.g.*, [31]).

In the remainder of this section we allow an arbitrary space dimension n; the exposition here follows [45]. With our signature $(-+\cdots+)$ one needs to repeat the calculations of Sect. 5.1 with the Hilbert Lagrangian density $\tilde{L}$ replaced by

$$\frac{\sqrt{-\det g_{\mu\nu}}}{16\pi}\left(g^{\alpha\beta}R_{\alpha\beta} - 2\Lambda\right) \,,$$

and without making the assumption $n+1=4$ done there. The resulting equations are identical up to (5.8), while that last equation gets replaced by

[6] A warped product form of the metric, with the factor in front of h not being constant, together with the Einstein equations (5.71), force g_{rr} and g_{tt} to have the form (5.63) in an appropriate coordinate system [31], with k being a function of r which approaches a constant as r tends to infinity. Further h itself has to satisfy the Einstein equation (5.71) with Λ replaced by an appropriate constant.

$$\Gamma^\alpha_{\mu\nu} = B^\alpha_{\mu\nu} - p^\alpha_{\mu\nu} + \frac{2}{n}\delta^\alpha_{(\mu}p^\kappa_{\nu)\kappa} \, .$$

The calculation continues as before: subtracting (5.6) from (5.4) and using the definition of $p^\alpha_{\mu\nu}$ we arrive at the equation

$$\mathfrak{g}^{\mu\nu}R_{\mu\nu} - \frac{\sqrt{-\det g_{\mu\nu}}}{8\pi}\Lambda = -\partial_\alpha\left(\mathfrak{g}^{\mu\nu}p^\alpha_{\mu\nu}\right) + L \, ,$$

where the new Lagrangian L differs from the old one by the cosmological constant term:

$$L := \mathfrak{g}^{\mu\nu}\left[\left(\Gamma^\alpha_{\sigma\mu} - B^\alpha_{\sigma\mu}\right)\left(\Gamma^\sigma_{\alpha\nu} - B^\sigma_{\alpha\nu}\right) - \left(\Gamma^\alpha_{\mu\nu} - B^\alpha_{\mu\nu}\right)\left(\Gamma^\sigma_{\alpha\sigma} - B^\sigma_{\alpha\sigma}\right) + \mathring{R}_{\mu\nu}\right]$$
$$- \frac{\sqrt{-\det g_{\mu\nu}}}{8\pi}\Lambda \, .$$

Similarily as in (5.10) we may calculate $p^\alpha_{\mu\nu}$ in terms of the derivatives of $\mathfrak{g}^{\mu\nu}$. The final result is:

$$p^\lambda_{\mu\nu} = \frac{1}{2}\mathfrak{g}_{\mu\alpha}\mathfrak{g}^{\lambda\alpha}{}_{;\nu} + \frac{1}{2}\mathfrak{g}_{\nu\alpha}\mathfrak{g}^{\lambda\alpha}{}_{;\mu} - \frac{1}{2}\mathfrak{g}^{\lambda\alpha}\mathfrak{g}_{\sigma\mu}\mathfrak{g}_{\rho\nu}\mathfrak{g}^{\sigma\rho}{}_{;\alpha}$$
$$+ \frac{1}{2(n-1)}\mathfrak{g}^{\lambda\alpha}\mathfrak{g}_{\mu\nu}\mathfrak{g}_{\sigma\rho}\mathfrak{g}^{\sigma\rho}{}_{;\alpha} \, , \tag{5.67}$$

and we have assumed that $n \geq 2$. Further,

$$\Gamma^\alpha_{\mu\nu} - B^\alpha_{\mu\nu} = -p^\alpha_{\mu\nu} + \frac{1}{n-1}\mathfrak{g}_{\sigma\rho}\mathfrak{g}^{\sigma\rho}{}_{;(\mu}\delta^\alpha_{\nu)} \, .$$

We have

$$\frac{\partial L}{\partial \mathfrak{g}^{\mu\nu}{}_{,\lambda}} = \frac{\partial L}{\partial \mathfrak{g}^{\mu\nu}{}_{;\lambda}} = \frac{\partial L}{\partial \Gamma^\alpha_{\beta\gamma}}\frac{\partial \Gamma^\alpha_{\beta\gamma}}{\partial \mathfrak{g}^{\mu\nu}{}_{;\lambda}} = p^\lambda_{\mu\nu} \, , \tag{5.68}$$

with the last equality being obtained by tedious but otherwise straightforward algebra. It follows again that the tensor $p^\lambda_{\mu\nu}$ is the momentum canonically conjugate to the contravariant tensor density $\mathfrak{g}^{\mu\nu}$. Alternatively, one can calculate

$$L = \frac{1}{2}\mathfrak{g}_{\mu\alpha}\mathfrak{g}^{\mu\nu}{}_{;\lambda}\mathfrak{g}^{\lambda\alpha}{}_{;\nu} - \frac{1}{4}\mathfrak{g}^{\lambda\alpha}\mathfrak{g}_{\sigma\mu}\mathfrak{g}_{\rho\nu}\mathfrak{g}^{\mu\nu}{}_{;\lambda}\mathfrak{g}^{\sigma\rho}{}_{;\alpha}$$
$$+ \frac{1}{8}\mathfrak{g}^{\lambda\alpha}\mathfrak{g}_{\mu\nu}\mathfrak{g}^{\mu\nu}{}_{;\lambda}\mathfrak{g}_{\sigma\rho}\mathfrak{g}^{\sigma\rho}{}_{;\alpha} + \mathfrak{g}^{\mu\nu}\mathring{R}_{\mu\nu} - \frac{\sqrt{-\det g_{\mu\nu}}}{8\pi}\Lambda \, . \tag{5.69}$$

and check directly that

$$\frac{\partial L}{\partial \mathfrak{g}^{\mu\nu}{}_{;\lambda}} = p^\lambda_{\mu\nu} \, . \tag{5.70}$$

The geometric context for our remaining considerations in this section is the following: Let $\mathscr{S}$ be an n-dimensional spacelike hypersurface in a $n+1$-dimensional Lorentzian space-time (M, g), $n \geq 2$. Suppose that M contains an open set $\mathscr{U}$ with a global time coordinate t (with range not necessarily equal to $\mathbb{R}$), as well as a global "radial" coordinate $r \in [R,\infty)$, leading to local coordinate systems $(x^\mu) = (t,r,v^A)$, with (v^A) — local coordinates on some compact $n-1$ dimensional manifold $\check{M}$. We further require that $\mathscr{S}_{\rm ext} \equiv \mathscr{S} \cap \mathscr{U} = \{t=0\}$. (The symbols x^0 and t are used interchangeably in this section.) In other words, $\mathscr{U}$ is a space-time neighbourhood — perhaps small — of the hypersurface

$$\mathscr{S}_{\rm ext} \approx [R,\infty) \times \check{M} \ .$$

Assume that the metric g approaches (as r tends to infinity, in a sense which will be made precise shortly) a background metric b of the form (5.63). Suppose further that g satisfies the vacuum Einstein equations with a cosmological constant

$$R_{\mu\nu} - \frac{g^{\alpha\beta}R_{\alpha\beta}}{2} g_{\mu\nu} = -\Lambda g_{\mu\nu} \ , \qquad \Lambda = -\frac{n(n-1)}{2\ell^2} \ , \tag{5.71}$$

similarly for b. (The existence of a large family of such g's follows from the work in [62, 88].) Let

$$\theta^0 = \sqrt{\frac{r^2}{\ell^2}+k}\, dt \ , \quad \theta^1 = \frac{1}{\sqrt{\frac{r^2}{\ell^2}+k}} dr \ , \quad \theta^P = r\alpha^P \ , \tag{5.72}$$

where α^P is an h-orthonormal coframe (the indices A, B, C run from 2 to n and are coordinate indices on $\check{M}$, while the indices P, Q, R have the same range and are tetrad indices there). We let e_a be the frame dual to θ^a,

$$e_0 = \frac{1}{\sqrt{\frac{r^2}{\ell^2}+k}} \partial_t \ , \quad e_1 = \sqrt{\frac{r^2}{\ell^2}+k}\, \partial_r \ , \quad e_P = \frac{1}{r}\beta_P \ , \tag{5.73}$$

so that β_P is a h-orthonormal frame on $(\check{M}, h)$. We set

$$e^{\mu\nu} \equiv g^{\mu\nu} - b^{\mu\nu} \ , \tag{5.74}$$

and let

$$e^{ab} \equiv g(\theta^a, \theta^b) - \eta^{ab}$$

denote the coefficients of $e^{\mu\nu}$ with respect to the frame θ^a dual to the e_a's:

$$e^{\mu\nu}\partial_\mu \otimes \partial_\nu = e^{ac} e_a \otimes e_c \ .$$

Here $\eta^{ab} = \mathrm{diag}(-1,+1,\cdots,+1)$. Thus — in this section — lower case Latin indices a, b, *etc.*, range from 0 to n and are tetrad indices, while Greek indices α, β, *etc.*, have the same ranges but are coordinate indices. We stress that we do *not* assume existence of global frames on the asymptotic region: when

$\breve{M}$ is not parallelizable, then any conditions on the e^{ab}'s, *etc.* assumed below should be understood as the requirement of *existence of a covering of* $\breve{M}$ *by a finite number of open sets* $\mathcal{O}_i$ *together with frames defined on* $[R_0,\infty)\times\mathcal{O}_i$ *of the form given above.*

We define the *phase space* $\mathscr{P}_{\mathrm{ads}}$ as the space of those smooth sections $(p^\lambda_{\mu\nu},\mathfrak{g}^{\alpha\beta})$ along $\mathscr{S}$ of the space-time phase bundle which satisfy the following conditions:

$\mathscr{C}1$. First, we only allow those sections of the space-time phase bundle which arise from solutions of the vacuum Einstein equations with cosmological constant Λ — in particular the general relativistic constraint equations with cosmological constant Λ have to be satisfied by the fields $(p^\lambda_{\mu\nu},\mathfrak{g}^{\alpha\beta})$.

$\mathscr{C}2$. Next, the e_a-tetrad components of g are required to be bounded on $\mathscr{S}_{\mathrm{ext}}$. Moreover, we impose the integral condition

$$\int_{\mathscr{S}_{\mathrm{ext}}} r\left(\sum_{a,b,c}|\mathring{\nabla}_a e^{bc}|^2+\sum_{d,e}|e^{de}|^2\right)d\mu_b<\infty\,, \tag{5.75}$$

where $\mathring{\nabla}$ is the Levi-Civita connection of the background metric b. In Equation (5.75) $d\mu_b$ is the measure arising from the metric induced on $\mathscr{S}$ by the background metric b; in local coordinates such that $\mathscr{S}_{\mathrm{ext}}=\{x^0=0\}$ we have $d\mu_b=\sqrt{\det b_{ij}}\,dr\,d^{n-1}v$, with the indices i,j running from 1 to n.

$\mathscr{C}3$. Further, the fall-off conditions

$$e^{ab}=o(r^{-n/2})\,,\quad e_c(e^{ab})=o(r^{-n/2})\,, \tag{5.76}$$

are assumed to hold on $\mathscr{S}_{\mathrm{ext}}$.

$\mathscr{C}4$. Finally, we shall assume that the following "volume normalisation condition" is satisfied:

$$\int_{\mathscr{S}_{\mathrm{ext}}} r|b_{cd}e^{cd}|\,d\mu_b<\infty\,. \tag{5.77}$$

Whenever we consider variations $(\delta p^\lambda_{\mu\nu},\delta\mathfrak{g}^{\alpha\beta})$ of the fields in $\mathscr{P}_{\mathrm{adS}}$, we will require that those variations satisfy the same decay conditions as the fields in $\mathscr{P}_{\mathrm{adS}}$.

In this section we will ignore the question of the choice of gauge maps in (5.44), assuming that some such choice has been made, compatible with the conditions listed above.

From now on we shall assume that $B^\alpha_{\beta\gamma}$ is the Levi-Civita connection of the background metric b. A condition equivalent to (5.75), which "looks less geometric" but is slightly more convenient to work with, is

$$\int_{\mathscr{S}_{\mathrm{ext}}} r\left(\sum_{a,b,c}|e_a(e^{bc})|^2+\sum_{d,e}|e^{de}|^2\right)d\mu_b<\infty\,. \tag{5.78}$$

This equivalence can be seen as follows: Let θ^i, e_i be as in Equations (5.72)-(5.73), we denote by

$$\mathring{\omega}^i{}_{jk} \equiv \theta^i(\mathring{\nabla}_{e_k} e_j)$$

the associated connection coefficients. One easily finds

$$\mathring{\omega}^P{}_{1Q} = \frac{\sqrt{\frac{r^2}{\ell^2}+k}}{r}\delta^P_Q = -\mathring{\omega}^1{}_{PQ} \, . \tag{5.79}$$

If we denote by

$$\alpha^P \equiv \alpha(v^A)^P{}_B dv^B \tag{5.80}$$

an orthonormal frame for the metric h, and by $\beta^P{}_{QR}$ the associated Levi-Civita connection coefficients, then

$$\mathring{\omega}^P{}_{QR} = \frac{1}{r}\beta^P{}_{QR} \, . \tag{5.81}$$

All connection coefficients other than those in (5.79) or (5.81) vanish. Equations (5.79)-(5.81) show that the $\mathring{\nabla}$-connection coefficients are bounded in the frame e_a, and the equivalence of (5.75) with (5.78) easily follows.

It is instructive to give some simpler fall-off conditions that ensure that $\mathscr{C}2$–$\mathscr{C}4$ hold; one possibility is

$$e^{ab} = O(r^{-\beta}) \, , \quad e_a(e^{bc}) = O(r^{-\beta}) \, , \quad b_{ab}e^{ab} = O(r^{-\gamma}) \, , \tag{5.82}$$

with

$$\beta > n/2 \, , \qquad \gamma > n \, . \tag{5.83}$$

We wish to verify that all the objects at hand are well defined on $\mathscr{P}_{\mathrm{adS}}$; let us start by showing that Equation (5.75) guarantees that the integral (5.14) defining $\Omega_{\mathscr{S}}$ converges. In order to see that, consider the identity:

$$\begin{aligned}\mathring{\nabla}_\alpha e^{\mu\nu} &= \mathring{\nabla}_\alpha g^{\mu\nu} = (\mathring{\nabla}_\alpha - \nabla_\alpha) g^{\mu\nu} \\ &= -g^{\mu\sigma}(\Gamma^\nu_{\sigma\alpha} - B^\nu_{\sigma\alpha}) - g^{\nu\sigma}(\Gamma^\mu_{\sigma\alpha} - B^\mu_{\sigma\alpha}) \, .\end{aligned}$$

The usual cyclic permutations calculation allows one to express $\Gamma^\alpha_{\beta\gamma} - B^\alpha_{\beta\gamma}$ as a linear combination of the $\mathring{\nabla}_\alpha e^{\mu\nu}$'s. It then follows from Equation (5.7) that the tetrad coefficients p^a_{bc} of $p^\alpha_{\beta\gamma}$ are, on $\mathscr{S}_{\mathrm{ext}}$, linear combinations with bounded coefficients of the $\mathring{\nabla}_a e^{bc}$'s. In local coordinates on $\mathscr{S}_{\mathrm{ext}}$ we have

$$\sqrt{|\det b_{\mu\nu}|} \sim r\sqrt{\det b_{ij}} \, ,$$

hence

$$\begin{aligned}\int_{\mathscr{S}_{\mathrm{ext}}} |\delta p^0_{ab}|\, |\delta \mathfrak{g}^{ab}| dr d^{n-1}v &\le C \sum_{a,b,c,d,e} \int_{\mathscr{S}_{\mathrm{ext}}} r|\mathring{\nabla}_c \delta e^{de}|\, |\delta e^{ab}| d\mu_b \\ &\le \frac{C}{2} \sum_{a,b,c,d,e} \int_{\mathscr{S}_{\mathrm{ext}}} r\left(|\mathring{\nabla}_c \delta e^{de}|^2 + |\delta e^{ab}|^2\right) d\mu_b < \infty \, .\end{aligned}$$

Here the coordinate x^0 has been chosen so that $\mathscr{S}_{\rm ext} = \{x^0 = 0\}$. Thus, $\Omega_{\mathscr{S}}$ is well defined on $\mathscr{P}_{\rm adS}$, as desired.

Recall, next, the ADM form Equation (5.33); similar considerations lead to the conclusion that the volume integral in Equation (5.33) converges. We wish to show that the boundary integral in Equation (5.33) vanishes for fields in $\mathscr{P}_{\rm adS}$, so that $\Omega_{\mathscr{S}}$ coincides with the more familiar "ADM symplectic form". We have

$$\begin{aligned}\delta\left(\frac{\sqrt{\det g_{ij}}}{N}\right) &= \delta\left(\sqrt{|\det g_{\mu\nu}|}\right) \\ &= \delta\left(\sqrt{\frac{\det g_{\mu\nu}}{\det b_{\mu\nu}}}\right)\sqrt{|\det b_{\mu\nu}|} \\ &= o(r^{-n/2})O(r^{n-1}) = o(r^{n/2-1})\ .\end{aligned}$$

To avoid a clash of notation with frame indices, the quantity N^3 appearing in Equation (5.33) will be denoted by N^r, g^{00} appearing there will be denoted by g^{tt}, *etc.* One easily checks the identity

$$N^r = \frac{n(r)}{\sqrt{|g^{tt}|}}\ ,$$

where n is the future directed g-unit-normal to $\mathscr{S}$. We have

$$\begin{aligned}g^{tt} \equiv g(dt,dt) &= (\eta^{ab}+e^{ab})e_a(t)e_b(t) \\ &= (\eta^{00}+e^{00})\left(e_0(t)\right)^2 \\ &= (\eta^{00}+e^{00})|b^{tt}|\ ,\end{aligned}$$

which gives

$$\frac{1}{\sqrt{|g^{tt}|}} = O(r)\ ,$$

$$\delta\left(\frac{1}{\sqrt{|g^{tt}|}}\right) = \delta\left(\sqrt{\frac{b^{tt}}{g^{tt}}}\right)\frac{1}{\sqrt{|b^{tt}|}} = o(r^{-n/2+1})\ .$$

Further,

$$\begin{aligned}n(r) = n^b e_b(r) = n^1 e_1(r) &= o(r^{-n/2+1})\ , \\ \delta n(r) = \delta n^1 e_1(r) &= o(r^{-n/2+1})\ ,\end{aligned}$$

where e_a is a b-orthonormal frame as in Equation (5.73), and the vanishing of the boundary term in (5.33) readily follows.

We have shown in the preceding sections that the Hamiltonian associated with a one parameter family of maps of the phase space into itself which arise from the flow of a vector field X on the space-time equals

$$H(X,\mathscr{S}) = \int_{\mathscr{S}} (p^{\mu}_{\alpha\beta}\pounds_X \mathfrak{g}^{\alpha\beta} - X^{\mu}L)\, dS_{\mu} \tag{5.84}$$

provided that all the integrals involved are well defined, and that the boundary integral in the variational formula

$$\begin{aligned} -\delta H = &\int_{\mathscr{S}} \left(\pounds_X p^{\lambda}{}_{\mu\nu}\delta\mathfrak{g}^{\mu\nu} - \pounds_X \mathfrak{g}^{\mu\nu}\delta p^{\lambda}{}_{\mu\nu}\right) dS_{\lambda} \\ &+ \int_{\partial\mathscr{S}} X^{[\mu}p^{\nu]}{}_{\alpha\beta}\delta\mathfrak{g}^{\alpha\beta}\, dS_{\mu\nu} \ , \end{aligned} \tag{5.85}$$

vanishes. Let us start with the question of convergence of H given by (5.84). First, one can check that all the formulae of [38, Appendix B] are dimension independent, and for vector fields X which are Killing vector fields for the metric b one has the identity

$$H^{\lambda} := p^{\lambda}{}_{\alpha\beta}\pounds_X \mathfrak{g}^{\alpha\beta} - X^{\lambda}L = \mathbb{W}^{\lambda\nu}{}_{;\nu} + T^{\lambda}{}_{\kappa}X^{\kappa} \ , \tag{5.86}$$

where $T^{\lambda}{}_{\kappa}$ is the matter energy-momentum tensor (assumed to vanish here), and $\mathbb{W}^{\nu\lambda}$ is given by Equation (5.17). If b is the anti-de Sitter metric, then the integral of $H^{\lambda}dS_{\lambda}$ over large "balls" $B_R := \{r \le R\}$ within $\mathscr{S}$ diverges if we try to pass with the radius of those balls to infinity. Indeed, we have

$$H^{\lambda}\Big|_{g=b} = -(\mathring{R} - 2\Lambda)X^{\lambda}/16\pi \ ,$$

with $\mathring{R}$ — the Ricci scalar of the background metric b, and $\mathring{R} - 2\Lambda = 4\Lambda/(n-1)$ in an $(n+1)$-dimensional space-time. We therefore subtract from H^{λ} a g-independent term which will cancel this divergence: indeed, such terms can be freely added to the Hamiltonian because they do not affect the variational formula (5.85) that defines it; from an energy point of view such an subtraction corresponds to a choice of the zero point of the energy. We thus set

$$\mathbb{U}^{\alpha\beta} := \mathbb{W}^{\alpha\beta}[g] - \mathbb{W}^{\alpha\beta}[b] \ ,$$

and from what has been said it follows that

$$m(\mathscr{S}, g, b, X) = \frac{1}{2}\int_{\partial\mathscr{S}} \mathbb{U}^{\alpha\beta} dS_{\alpha\beta} \ , \tag{5.87}$$

will still be a candidate for a Hamiltonian. From the definition of H^{λ} and from Equation (5.86) one easily finds

$$\begin{aligned} 16\pi\mathring{\nabla}_{\beta}\mathbb{U}^{\alpha\beta} = &\left(\sqrt{|\det g|}g^{ab} - \sqrt{|\det b|}b^{ab}\right)\mathring{R}_{ab}X^{\beta} \\ &+ 2\Lambda\left(\sqrt{|\det b|} - \sqrt{|\det g|}\right)X^{\beta} \\ &+ \sqrt{|\det b|}\left(Q^{\alpha}{}_{\beta}X^{\beta} + Q^{\alpha}{}_{\beta\gamma}\mathring{\nabla}^{\beta}X^{\gamma}\right) \ , \end{aligned} \tag{5.88}$$

where $Q^\alpha{}_\beta$ is a quadratic form in $e_a(e^{bc})$, and $Q^\alpha{}_{\beta\gamma}$ is bilinear in $e_a(e^{bc})$ and e^{ab}, both with bounded coefficients. In order to analyze convergence of the integral of H^λ we need some hypotheses on the vector field X. We maintain of course the condition that X is a b-Killing vector field. In the current context it is then natural to suppose that

$$|X| + |\mathring{\nabla} X| \leq Cr \tag{5.89}$$

for some constant C. This is easily seen to hold for the Killing vector field $X = \partial/\partial t$, using Equations (5.79) and (5.81). Similarly one can show that (5.89) is satisfied by all Killing vectors of the $n+1$-dimensional anti-de Sitter metrics, see [45] for details.

We are finally ready to show the convergence of the Hamiltonians (5.87). We have

$$\int_{\{r=R\}} \mathbb{U}^{\alpha\beta} dS_{\alpha\beta} = 2\int_{\{R_0\leq r\leq R\}} \mathring{\nabla}_\beta \mathbb{U}^{\alpha\beta} dS_\alpha + \int_{\{r=R_0\}} \mathbb{U}^{\alpha\beta} dS_{\alpha\beta} \,. \tag{5.90}$$

Formula (5.88) for the volume integrand in Equation (5.90) clearly guarantees convergence of that volume integral to a finite value when R tends to infinity for all fields in $\mathscr{P}_{\mathrm{adS}}$.

As our last restriction, we shall suppose that the b-Killing vector field X has the property that *the associated variations of the fields are compatible with the boundary conditions* imposed on fields in $\mathscr{P}_{\mathrm{adS}}$. This means in particular that we must have

$$\int_{\mathscr{S}} r \sum_{a,b,c} |\pounds_X \left(\mathring{\nabla}_a e^{bc}\right)|^2 \, d\mu_b < \infty \,. \tag{5.91}$$

Clearly the volume integral in the variational formula (5.85) converges under (5.91), together with the remaining conditions set forth above. Further, the boundary integral there vanishes under (5.76). We have therefore shown that — under the above conditions on X — (5.87) provides the required Hamiltonian on $\mathscr{P}_{\mathrm{adS}}$.

It can be checked (*cf.* [45, Appendix C]) that for fields in $\mathscr{P}_{\mathrm{adS}}$ the numerical value of (5.87) with $X = \partial_t$ coincides with the Abbott-Deser mass, introduced in [1] by completely different methods.

The reader is referred to [45] for an analysis of the geometric character of the global charges defined via the integrals (5.87).

In the current context there exist several alternative methods for defining mass [10, 11, 14, 27, 68, 69, 99]; an extended discussion can be found in [46, Section 5]. An alternative Hamiltonian analysis is carried out in [75]. We wish to stress that the key advantage of the Hamiltonian approach is the uniqueness of the candidate expression for the energy (which follows from the fact that Hamiltonians are uniquely defined up to a constant on each path connected component of the phase space), and that no such uniqueness properties are known in the alternative approaches mentioned above (*cf.*, however [87, 114, 126]).

We end this section by noting the following simple form of (5.87), derived in [46], which holds for gravitational fields in $\mathscr{P}_{\mathrm{adS}}$ in dimension 3 + 1 when the Killing vector field X equals $\partial/\partial t$:

$$m(\mathscr{S}, g, b, \partial_t) = \lim_{R\to\infty} \frac{R^3}{16\pi\ell^2} \int_{\Sigma\cap\{r=R\}} \left(r \sum_{P=2}^{n} \frac{\partial e^{PP}}{\partial r} - 2e^{11} \right) d^2\mu_h \,, \tag{5.92}$$

where $d^2\mu_h$ is the Riemannian measure associated with the metric h on $\breve{M}$.

5.6 Energy in the radiation regime: convergence of integrals

Let us return to the main point of our work here, namely the Hamiltonian analysis of the gravitational field in the radiation regime. Our analysis will run closely in parallel to that of the radiating scalar field. The hypersurfaces $\mathscr{S}$ we shall consider from now on will be the general relativistic analogues of hyperbolae in Minkowski space-time, the "asymptotically hyperboloidal" hypersurfaces to be defined precisely below. As in Chap. 4 the model hypersurface $\widetilde{\Sigma}$ — the equivalent of the hypersurface Σ of Chap. 3 — will consist of two parts, the first, Σ, being mapped by ψ to the space-time M, and the second, $\Sigma_{\mathscr{I}^+}$, being mapped by ψ to the boundary of M in its conformal completion $\widetilde{M}$, *cf.* equation (4.9). As in Chap. 4 the dynamics of that part of $\widetilde{\Sigma}$ which lies in M will be by moving $\psi_{I\times\Sigma}(\Sigma)$ along the flow of a vector field X defined on M; here $\psi_{I\times\Sigma}$ is as in equation (4.11). To make things precise, let us start by defining the conformal completion $\widetilde{M} = M \cup \mathscr{I}^+$ of M. We shall suppose that there exists an open subset of M on which Bondi–Sachs coordinates can be introduced:

$$g_{\mu\nu}\mathrm{d}x^\mu \mathrm{d}x^\nu = -\frac{V}{r}\mathrm{e}^{2\beta}\mathrm{d}u^2 - 2\mathrm{e}^{2\beta}\mathrm{d}u\mathrm{d}r + r^2 h_{AB}(\mathrm{d}x^A - U^A\mathrm{d}u)(\mathrm{d}x^B - U^B\mathrm{d}u) \tag{5.93}$$

$$\frac{\partial(\det h_{AB})}{\partial r} = 0 \,, \tag{5.94}$$

$$r \in (r_0, \infty) \,, \quad u \in (u_-, u_+) \,, \quad (x^A) \text{ — local coordinates on } S^2 \,.$$

We replace the area coordinate r by its inverse,

$$x = 1/r \in (0, 1/r_0) \,,$$

and will use the coordinate system (u, x^A, x) to describe the field in a neighbourhood of $\mathscr{I}^+$. (Although our constructions are analogous to the constructions done for the scalar field in Chap. 4, the coordinate x used here increases towards the interior of space-time, in contrast to the coordinate ρ used there. As a result, the natural exterior orientation of $\mathscr{I}^+$ is opposite to the orientation carried by dx.)

We shall require that in the coordinate system (u, x^A, x) the functions V, β, the S^2-vector fields U^A, and the S^2-Riemannian metrics h_{AB} extend smoothly by continuity to the boundary

$$\mathscr{I}^+ \equiv \{x = 0\} .$$

Finally we shall require that

$$\lim_{x \to 0} h_{AB} = \breve{h}_{AB} ,$$

where $\breve{h}_{AB}$ is the standard round metric on the sphere,

$$\breve{h}_{AB} dx^A dx^B = d\theta^2 + \sin^2\theta \, d\varphi^2 .$$

We set

$$\widetilde{M} \equiv M \cup \mathscr{I}^+ ,$$

with the obvious differential structure defined by the coordinate system (u, x^A, x). This construction of $\mathscr{I}^+$ and $\widetilde{M}$ is equivalent [118] to the geometric approach of Penrose [104] as far as local considerations near $\mathscr{I}^+$ are concerned. Here one chooses the conformal factor Ω as

$$\Omega \equiv x ,$$

so that the unphysical metric $\tilde{g}$ is

$$\tilde{g} \equiv x^2 g .$$

The question of existence of vacuum space-times satisfying the above conditions is discussed in Appendix C.2.

A (connected) hypersurface $\mathscr{S} \subset \widetilde{M}$ will be said to be *asymptotically hyperboloidal (or shortly hyperboloidal) with compact interior* if the complement of a compact subset of $\mathscr{S}$ can be written as

$$\mathscr{S}_{\text{ext}} = \{u = \alpha(x^A, x)\} , \tag{5.95}$$

in the coordinate system above, for some smooth[7] function α. Further we shall require that, in the local coordinates (u, x, x^A), the metric induced by $\tilde{g}$ on the hypersurface $\mathscr{S}$ be uniformly positive definite up-to-boundary. It is straightforward to extend our analysis to the case in which $\mathscr{S}$ has a finite number of asymptotic regions as above, the details are left to the reader. An initial data set (Σ, γ, K) will be called *asymptotically hyperboloidal (or shortly*

[7] In Sect. 6.10, where polyhomogeneous metrics are considered, we will allow functions α which are polyhomogeneous.

hyperboloidal) with compact interior if there exists an embedding i of Σ into some space-time (M, g) such that $i(\Sigma)$ satisfies the conditions spelled out above. *In the remainder of this paper the symbol $\mathscr{S}$ stands for hypersurfaces which are asymptotically hyperboloidal with compact interior.* We note that with our definition here the asymptotic behaviour of the extrinsic curvature tensor field K is strongly constrained by conditions which follow implicitly from the requirements set forth above, together with the differentiability conditions — smoothness or polyhomogeneity — on the functions which appear in the Bondi form of the metric, assumed in this work.

Let us pass now to the question of convergence of the integral appearing in $\Omega_{\mathscr{S}}$, as defined by equation (5.14). As already pointed out, the background connection drops out from the variations of the fields, which appear in equation (5.14), hence the convergence of the integral defining $\Omega_{\mathscr{S}}$ — or lack thereof — does not depend upon the choice of the background. (We will soon choose the background connection to be, *in a neighborhood*[8] *of* $\mathscr{I}^+$, the Levi–Civita connection of the flat metric

$$b = b_{\mu\nu} dx^\mu dx^\nu \equiv -du^2 - 2du\, dr + r^2 \breve{h}_{AB} dx^A dx^B \ , \tag{5.96}$$

but this is irrelevant for $\Omega_{\mathscr{S}}$. Here and throughout $\breve{h}_{AB}\mathrm{d}x^A\mathrm{d}x^B = \mathrm{d}\theta^2 + \sin^2\theta\, \mathrm{d}\varphi^2$. The form of the metric (5.93) leads to the following equations:

$$\delta\mathfrak{g}^{03} = \delta\mathfrak{g}^{0A} = \delta\mathfrak{g}^{00} = \delta p^\mu{}_{33} = \delta p^0{}_{3A} = 0 \tag{5.97}$$

(see Appendix C.1, *cf.* also [83, Section 4]). We shall also need the following expansions for the coefficients of the metric (the formulae below are taken from [43]; they are essentially due to [123], *cf.* also [25])

$$h_{AB} = \breve{h}_{AB}\left(1 + \frac{1}{4r^2}\chi^{CD}\chi_{CD}\right) + \frac{\chi_{AB}(v)}{r} + O(r^{-3}) \ , \tag{5.98}$$

$$\beta = -\frac{\breve{h}^{AB}\breve{h}^{CD}\chi_{AC}\chi_{BD}}{32r^2} + O(r^{-3}) \ ,$$

$$\begin{aligned} U^A &= -\frac{\breve{\mathscr{D}}_B\chi^{AB}}{2r^2} + \frac{2N^A(v)}{r^3} + \frac{\breve{\mathscr{D}}^A\left(\chi^{CD}\chi_{CD}\right)}{16r^3} \\ &\quad + \frac{\chi^A{}_B\breve{\mathscr{D}}_C\chi^{BC}}{2r^3} + O(r^{-4}) \ , \end{aligned} \tag{5.99}$$

$$\begin{aligned} V &= r - 2M(v) + \frac{\breve{\mathscr{D}}_B\chi^{AB}\breve{\mathscr{D}}^C\chi_{AC} - 4\breve{\mathscr{D}}_A N^A}{4r} \\ &\quad + \frac{\chi^{CD}\chi_{CD}}{16r} + O(r^{-2}) \ . \end{aligned} \tag{5.100}$$

[8] It should be pointed out that we do not assume that the background metric is necessarily of the form (5.96) throughout the manifold (which would impose undesirable global constraints on the manifold M) — (5.96) is assumed to hold only in the asymptotic region. In particular, $B^\sigma{}_{\beta\gamma\kappa} = 0$ does not have to vanish throughout M; under our conditions on $\mathscr{S}$ the $B^\sigma{}_{\beta\gamma\kappa}$'s will, however, be compactly supported.

Here $v \equiv (u, x^A)$, and $\breve{\mathscr{D}}_A$ is the covariant derivative operator associated with the metric $\breve{h}_{AB}$ on S^2. Indices A, B, *etc.*, take values 1 and 2, and are raised and lowered with $\breve{h}^{AB}$. The tensor field χ_{AB} is trace-free with respect to the metric $\breve{h}$:

$$\breve{h}^{AB}\chi_{AB} = 0 . \tag{5.101}$$

Further, the functions M and N^A satisfy the following evolution equations

$$\frac{\partial M}{\partial u} = -\frac{1}{8}\breve{h}^{AC}\breve{h}^{BD}\partial_u\chi_{AB}\partial_u\chi_{CD} + \frac{1}{4}\breve{\mathscr{D}}_A\breve{\mathscr{D}}_B\partial_u\chi^{AB} , \tag{5.102}$$

$$3\frac{\partial N^A}{\partial u} = -\breve{\mathscr{D}}^A M + \frac{1}{4}\epsilon^{AB}\breve{\mathscr{D}}_B\tilde{\lambda} - K^A , \tag{5.103}$$

$$K^A \equiv \frac{3}{4}\chi^A{}_B\breve{\mathscr{D}}_C\partial_u\chi^{BC} + \frac{1}{4}\partial_u\chi^{CD}\breve{\mathscr{D}}_D\chi^A{}_C ,$$

$$\tilde{\lambda} \equiv \breve{h}^{BD}\epsilon^{AC}\breve{\mathscr{D}}_C\breve{\mathscr{D}}_B\chi_{DA} .$$

Here ϵ_{AB} is a tensor field on S^2 defined by the formula $\mathrm{d}^2\mu \equiv \sin\theta\,\mathrm{d}\theta\wedge\mathrm{d}\varphi = \frac{1}{2}\epsilon_{AB}dx^A\wedge dx^B$. If we fix some $u_0 \in I$, then the Einstein equations do not impose any local restrictions on the function $M(u_0, \theta, \varphi)$, and on the vector field $N^A(u_0, \theta, \varphi)$ on S^2, or on the tensor field $\chi_{CD}(u, \theta, \varphi)$ on S^2. Inserting the expansions (5.98)–(5.100) into (C.9)–(C.25), in addition to (5.97) one finds

$$16\pi\delta\mathfrak{g}^{AB} = -x^{-1}\sin\theta\,\delta\chi^{AB} + O(1) , \tag{5.104}$$

$$\delta\mathfrak{g}^{33} = O(x) , \tag{5.105}$$

$$\delta\mathfrak{g}^{3A} = O(1) , \tag{5.106}$$

$$\delta p^A{}_{3B} = -\frac{1}{2}\delta\chi^A{}_B + O(x) , \tag{5.107}$$

$$\delta p^0{}_{AB} = -\frac{1}{2}\delta\chi_{AB} + O(x) , \tag{5.108}$$

$$\text{all other} \quad \delta p^\lambda{}_{\mu\nu} = O(x) . \tag{5.109}$$

We can now pass to the question of convergence of the integral

$$\int_{\mathscr{S}} (\delta_1 p^\mu_{\alpha\beta}\delta_2\mathfrak{g}^{\alpha\beta} - \delta_2 p^\mu_{\alpha\beta}\delta_1\mathfrak{g}^{\alpha\beta})\,dS_\mu \tag{5.110}$$

on spacelike hypersurfaces $\mathscr{S}$ of the form (5.95). To start with, we shall for simplicity assume that

$$\partial_A\alpha = 0 . \tag{5.111}$$

The convergence of the various relevant integrals without assuming (5.111) follows from the results in Sect. 6.1. equations (5.97) together with (5.104)–(5.109) show (*cf.* also equation (4.78)) that all terms appearing in (5.14) are $O(1)$ except for the term $\delta_1 p^0_{AB}\delta_2\mathfrak{g}^{AB} - \delta_2 p^0_{AB}\delta_1\mathfrak{g}^{AB}$. However, the explicit form of the $O(1/x)$ terms in $\delta\mathfrak{g}^{AB}$ together with equation (5.108) show that the dangerous terms cancel out, so that $\Omega_{\mathscr{S}}((\delta_1 p^\lambda_{\mu\nu}, \delta_1 \mathfrak{g}^{\alpha\beta}), (\delta_2 p^\lambda_{\mu\nu}, \delta_2 \mathfrak{g}^{\alpha\beta}))$ as defined in (5.14) is indeed well defined for metrics satisfying the expansions (5.98)–(5.100).

Let us pass now to the question of convergence of the Hamiltonian (5.15). We wish, first, to analyze Hamiltonians related to motions along vector fields X such that

$$X = \frac{\partial}{\partial u}$$

in the asymptotic region defined by the Bondi coordinate system. *From now on we define the background metric in the asymptotic region by equation (5.96), with $B^\alpha_{\beta\gamma}$ — the Levi–Civita connection of $b_{\beta\gamma}$.* We proceed similarly as in the case of the scalar field: we consider the compact hypersurfaces with boundary $\mathscr{S}^\epsilon$ obtained by removing from $\mathscr{S}$ the set $\{x \le \epsilon\}$. Equations (5.15)–(5.16) give

$$H(X, \mathscr{S}^\epsilon) = \frac{1}{2}\int_{\partial\mathscr{S}^\epsilon} \mathbb{W}^{\alpha\beta}\, dS_{\alpha\beta}\,. \tag{5.112}$$

For further purposes it is convenient to momentarily relax our assumption that $X = \partial/\partial u$; expanding out equations (5.17)–(5.18), one has

$$\begin{aligned}\mathbb{W}^{ux} = X^0 &\left[\mathfrak{g}^{30}(p^3{}_{30} - p^0{}_{00}) - \mathfrak{g}^{\mu\nu}p^3{}_{\mu\nu} - \mathfrak{g}^{3k}p^0{}_{k0} + \frac{1}{4}\mathfrak{g}^{3\mu}p^\nu{}_{\mu\nu}\right] \\ &+X^3\left[\mathfrak{g}^{\mu\nu}p^0{}_{\mu\nu} - \mathfrak{g}^{03}p^0{}_{03} - \frac{1}{3}\mathfrak{g}^{03}p^\sigma{}_{3\sigma}\right] + \mathfrak{g}^{\alpha 0}X^3{}_{;\alpha} - \mathfrak{g}^{\alpha 3}X^0{}_{;\alpha} \\ &+X^A\left[\mathfrak{g}^{30}(p^3{}_{3A} - p^0{}_{0A}) - \mathfrak{g}^{3B}p^0{}_{BA}\right]\end{aligned} \tag{5.113}$$

(here and everywhere below indices 0 and 3 refer to $x^0 = u$ and $x^3 = x$), where equations (C.9), (C.12)–(C.14), (C.20) and (C.8) have been used; here the index k runs over the range $1, 2, 3$. Using the formulae of Appendix C.1, after a long computation one obtains:

$$\begin{aligned}\mathbb{W}^{xu} = \frac{\sin\theta}{16\pi}&\Big\{X^u\left[x^{-2}\mathscr{D}_A U^A - 2V + x^{-1}e^{2\beta}h^{AB}\check{h}_{AB} + e^{-2\beta}x^{-2}U_AU^A{}_{,x}\right] \\ &-x^{-2}X^u{}_{,A}U^A + (1 - Vx)X^u{}_{,x} + x^{-2}X^x\left[2x^{-1} - x^{-1}e^{2\beta}h^{AB}\check{h}_{AB}\right] \\ &-x^{-2}X^A e^{-2\beta}h_{AB}U^B{}_{,x} + x^{-2}(X^x{}_{;x} - X^u{}_{;u}) - X^u{}_{;x}\Big\}\,,\end{aligned} \tag{5.114}$$

where $\mathscr{D}$ denotes the covariant derivative on S^2 associated with the metric h_{AB}; recall that a semi-column denotes covariant differentiation with respect to the background metric. Coming back to the case $X = \partial/\partial u$ (which is covariantly constant with respect to the background metric), from (5.111),

(5.114) and the asymptotics (5.98)–(5.100) of the field at $\mathscr{I}^+$ we obtain[9]

$$E_{TB}(\mathscr{S}) \equiv \lim_{\epsilon\to 0}\frac{1}{2}\int_{\partial\mathscr{S}^\epsilon} \mathbb{W}^{\alpha\beta}\, dS_{\alpha\beta} = \frac{1}{4\pi}\int_{S^2} M(u,\theta,\varphi)\sin\theta d\theta d\varphi\ , \quad (5.115)$$

a quantity which we call the Trautman–Bondi mass of the asymptotically hyperboloidal hypersurface $\mathscr{S}$ [25, 121]. (Recall that M is the coefficient of the expansion (5.100).)

To check the validity of the variational formula (5.21) for $\mathscr{S}$ — a hyperboloidal hypersurface with compact interior, we need again to establish the convergence of the integrals appearing there. Let us start by analyzing the volume integral in (5.21). As before we assume that $X = \partial/\partial u$, which is covariantly constant in the background (5.96), hence a Killing vector field thereof. An analysis of equations (C.9)–(C.8) yields the following asymptotic behaviour of the Lie derivatives that appear in (5.21):

$$\pounds_X {p^0}_{\mu\nu} = -\frac{1}{2}\partial_u \chi_{AB} + O(x)\ , \quad (5.116)$$

$$\pounds_X {p^k}_{\mu\nu} = O(x)\ , \quad (5.117)$$

$$16\pi \pounds_X \mathfrak{g}^{AB} = -x^{-1}\sin\theta\partial_u\chi^{AB} + O(1)\ . \quad (5.118)$$

The detailed behaviour of the remaining $\pounds_X\mathfrak{g}^{\mu\nu}$'s is irrelevant, as they are all $O(1)$ while the corresponding terms $\delta {p^\lambda}_{\mu\nu}$ are $O(x)$, which (more than) ensures a finite contribution to the integral (5.21). It follows from equations (5.97), (5.104)–(5.106) and (5.116)–(5.118) that the terms of order $1/x$ cancel again so that

$$\pounds_X\mathfrak{g}^{\mu\nu}\delta {p^\lambda}_{\mu\nu} - \pounds_X {p^\lambda}_{\mu\nu}\delta\mathfrak{g}^{\mu\nu} = O(1)\ ,$$

which together with an appropriate equivalent of (4.78) guarantees convergence of the volume integral in (5.21).

When $X = \partial/\partial u$ and the integral in (5.21) is taken over $\mathscr{S}^\epsilon \equiv \mathscr{S}\backslash\{x \le \epsilon\}$, the boundary term in (5.21) takes the form:

$$16\pi\left(X^0 {p^3}_{\mu\nu}\delta\mathfrak{g}^{\mu\nu} - X^3 {p^0}_{\mu\nu}\delta\mathfrak{g}^{\mu\nu}\right) = -\frac{1}{2}\sin\theta\partial_u\chi_{AB}\delta\chi^{AB} + O(\epsilon)\ .$$

It follows that we can safely pass to the limit $\epsilon = 0$ to obtain

$$-\delta H(X,\mathscr{S}) = \int_{\mathscr{S}} \left(\pounds_X {p^\lambda}_{\mu\nu}\delta\mathfrak{g}^{\mu\nu} - \pounds_X\mathfrak{g}^{\mu\nu}\delta {p^\lambda}_{\mu\nu}\right) dS_\lambda$$
$$+\frac{1}{32\pi}\int_{S^2} \sin\theta\partial_u\chi_{AB}\delta\chi^{AB}\, d\theta\, d\varphi\ , \quad (5.119)$$

[9] It has been shown in [43] that equation (5.115) remains true for all polyhomogeneous vacuum metrics. The reader is referred to Sect. 6.10 for a further discussion.

with a convergent volume integral, as established above (the sign in the boundary integral is determined by the exterior orientation of $\partial\mathscr{S}$, and the fact that the vector ∂_x is pointing towards the interior of $\mathscr{S}$).

It may be of some interest to enquire about convergence of integrals appearing in the ADM equivalent (5.43) of the variational formula (5.119). For simplicity of notation let us identify the hypersurfaces Σ_τ with their images $\mathscr{S}(\tau) \equiv \psi_\tau(\Sigma)$ in M under the map ψ of Chap. 3, see equation (3.5). We shall suppose that in a neighbourhood $\mathcal{O}$ of $\mathscr{I}^+$ coordinatized by Bondi coordinates the $\mathscr{S}_\tau$'s are of the form

$$\mathscr{S}(\tau) \cap \mathcal{O} = \{u = \alpha(x, x^A) + \tau\} ,$$

for some function α, which is smooth in all its arguments. Consider the identity (5.42) with $\Sigma = \mathscr{S}(\tau)^\epsilon \equiv \mathscr{S}(\tau) \setminus \{x \le \epsilon\}$; in the Bondi parameterization the fields μ and σ appearing in (5.42) can be calculated using equations (C.4)–(C.8), to yield

$$\begin{aligned}
\mu &= -16\pi \mathfrak{g}(du - d\alpha, du - d\alpha) \\
&= -x^{-2} \sin\theta \left[-2\alpha_x + Vx^3(\alpha_x)^2 + 2\alpha_x \alpha_A U^A + e^{2\beta} h^{AB} \alpha_A \alpha_B\right] , \\
\mu\sigma &= 16\pi \mathfrak{g}(du - d\alpha, dx) \\
&= x^{-2} \sin\theta \left(1 - \alpha_x V x^3 - \alpha_A U^A\right) .
\end{aligned} \tag{5.120}$$

It follows that

$$\delta(\mu\sigma) = O(1) , \qquad \frac{\partial(\mu\sigma)}{\partial\tau} = O(1) ,$$
$$\delta(\log\mu) = O(x) , \qquad \frac{\partial \log\mu}{\partial\tau} = O(x)$$

(we note that spacelikeness-up-to-$\mathscr{I}^+$ of $\mathscr{S}$ implies that μ has no zeros, so that $\log\mu$ is well defined up to $\mathscr{I}^+$), so that

$$\dot\sigma\delta\mu - \dot\mu\delta\sigma = (\dot{\sigma\mu})\delta\log\mu - (\dot{\log\mu})\delta(\sigma\mu) = O(x) . \tag{5.121}$$

This shows that the boundary integral in (5.42) vanishes in the limit $\epsilon \to 0$. In the case[10] of $\alpha|_{\mathscr{I}^+} = 0$ the left hand side of (5.42) has already been shown to converge when ϵ tends to zero, we thus obtain that the volume integral in the right hand side of (5.42) has a well defined limit when ϵ approaches zero. From equations (5.43) and (5.119) we obtain

$$\begin{aligned}
-\delta H(X, \mathscr{S}) = {} & \frac{1}{16\pi} \int_\Sigma \left(\dot\pi^{kl} \delta\gamma_{kl} - \dot\gamma_{kl} \delta\pi^{kl}\right) d^3x \\
& + \frac{1}{32\pi} \int_{S^2} \sin\theta \partial_u \chi_{AB} \delta\chi^{AB} \, d\theta \, d\varphi .
\end{aligned} \tag{5.122}$$

[10] The convergence analysis for more general α's is done in Sect. 6.1, so that the same argument applies to establish ADM type variational formulae in the general case.

A similar analysis may be performed for that part of $\tilde{\Sigma}$ which is mapped by ψ to the boundary $\mathscr{I}^+$ of M. The relevant field data will be described by the value on $\mathscr{I}^+$ of χ_{AB}. Passing to the limit as ϵ goes to zero on a family of hypersurfaces $x = \epsilon$ in equation (5.14) yields the following formula for $\Omega_{\mathscr{I}^+_I}$ on the space of functions χ_{AB}:

$$\begin{aligned}
\Omega_{\mathscr{I}^+_I}(\delta_1\chi, \delta_2\chi) &\equiv \lim_{\epsilon\to 0} \int_{\{x=\epsilon\}\times I\times S^2} (\delta_1 p^\mu{}_{\alpha\beta}\delta_2 \mathfrak{g}^{\alpha\beta} - \delta_2 p^\mu{}_{\alpha\beta}\delta_1 \mathfrak{g}^{\alpha\beta})\, dS_\mu \\
&= -\lim_{\epsilon\to 0}\int_{\tau\in I} d\tau \int_{S_\tau} (\delta_1 p^x{}_{\alpha\beta}\delta_2 \mathfrak{g}^{\alpha\beta} - \delta_2 p^x{}_{\alpha\beta}\delta_1 \mathfrak{g}^{\alpha\beta})\Big|_{x=\epsilon} \sin\theta\, d\theta\, d\varphi \\
&= \int_{\tau\in I} d\tau \int_{S_\tau} \left(\frac{\partial\delta_1\chi_{AB}}{\partial u}\delta_2\chi^{AB} - \frac{\partial\delta_2\chi_{AB}}{\partial u}\delta_1\chi^{AB}\right)\frac{\sin\theta}{32\pi}\, d\theta\, d\varphi \quad (5.123)
\end{aligned}$$

(the minus sign in the second integral above arises from exterior orientation considerations). We set

$$\begin{aligned}
H(X, \mathscr{I}^+_I) &\equiv -\lim_{\epsilon\to 0}\int_{\{x=\epsilon\}\times I\times S^2} H^x\, du\, d\theta\, d\varphi \\
&= -\int_I du \int_{S^2} \partial_u \mathbb{W}^{xu}\Big|_{\mathscr{I}^+} d\theta\, d\varphi = \int_{\partial\mathscr{I}^+_I} \mathbb{W}^{ux}\Big|_{\mathscr{I}^+} d\theta\, d\varphi\,. \quad (5.124)
\end{aligned}$$

In this context Equation (3.48) takes the following form

$$\begin{aligned}
-\delta H(X, \mathscr{I}^+_{[\tau_-,\tau]}) = \Omega_{\mathscr{I}^+_{[\tau_-,\tau]}}(\pounds_X\chi, \delta\chi) &- \frac{1}{32\pi}\int_{S_{\tau_-}} \frac{\partial\chi_{AB}}{\partial u}\delta\chi^{AB}\sin\theta\, d\theta\, d\varphi \\
&+ \frac{1}{32\pi}\int_{S_\tau} \frac{\partial\chi_{AB}}{\partial u}\delta\chi^{AB}\sin\theta\, d\theta\, d\varphi\,. \quad (5.125)
\end{aligned}$$

It is not immediately obvious that equation (5.124) leads to a Hamiltonian for the dynamics of the field at $\mathscr{I}^+$ — to see that this is actually the case one can use equations (5.124), (5.115) and (5.102) to obtain

$$\begin{aligned}
&H(X, \mathscr{I}^+_I) \\
&= -\lim_{\epsilon\to 0}\frac{1}{4\pi}\int_{\partial\mathscr{I}^+_I} M\sin\theta\, d\theta\, d\varphi \\
&= \frac{1}{16\pi}\int_{I\times S^2} \left(\frac{1}{2}\breve{h}^{AC}\breve{h}^{BD}\partial_u\chi_{AB}\partial_u\chi_{CD} - \breve{\mathscr{D}}_A\breve{\mathscr{D}}_B\partial_u\chi^{AB}\right)\sin\theta\, d\theta\, d\varphi \\
&= \frac{1}{32\pi}\int_{I\times S^2} \breve{h}^{AC}\breve{h}^{BD}\partial_u\chi_{AB}\partial_u\chi_{CD}\sin\theta\, d\theta\, d\varphi\,. \quad (5.126)
\end{aligned}$$

The above formula can also be established directly from the gravitational analogue of equation (4.38):

$$H^x\Big|_{\mathscr{I}^+} = p^x{}_{\mu\nu}\pounds_X\mathfrak{g}^{\mu\nu}\Big|_{\mathscr{I}^+} = p^x{}_{AB}\pounds_X\mathfrak{g}^{AB}\Big|_{\mathscr{I}^+} = -\frac{1}{32\pi}\sin\theta\partial_u\chi_{AB}\partial_u\chi^{AB}\,. \quad (5.127)$$

The direct derivation of the variational formula (5.125) for the scalar field, as carried out in equation (4.40), applies here with obvious modifications.

Gluing together $\partial\mathscr{S}_\tau$ with the corresponding piece of $\partial\mathscr{I}^+_{[\tau_-,\tau]}$, in a manner completely analogous to (4.41), we finally obtain:

$$-\delta H(X, \mathscr{S}_\tau \cup \mathscr{I}^+_{[\tau_-,\tau]}) = \Omega_{\mathscr{S}_\tau}((\pounds_X p, \pounds_X \mathfrak{g}), (\delta p, \delta\mathfrak{g})) + \Omega_{\mathscr{I}^+_{[\tau_-,\tau]}}(\pounds_X\chi, \delta\chi) + \frac{1}{32\pi}\int_{S_{\tau_-}} \frac{\partial\chi_{AB}}{\partial u}\delta\chi^{AB}\sin\theta\, d\theta\, d\varphi \quad (5.128)$$

with $\Omega_{\mathscr{S}_\tau}((\pounds_X p, \pounds_X \mathfrak{g}), (\delta p, \delta\mathfrak{g}))$ defined in (4.5) and $\Omega_{\mathscr{I}^+_{[\tau_-,\tau]}}(\pounds_X\chi, \delta\chi)$ defined by equation (5.123). The reader should note the similarity of this formula with the corresponding formula (4.41) for the scalar field.

5.7 Phase spaces: the space $\mathscr{P}$

The analysis in this section, as well as in the two following ones, runs largely in parallel with that in Sects. 4.3–4.5. There is, however, a major subtlety, which arises from the fact that *each general relativistic Cauchy data set*, say p, *leads to its own space-time* M_p, thus there is no single space-time M, which one can associate in a *canonical* way to all Cauchy data (*cf.* also [72, 97] for some related discussions). This is obviously related to the coordinate invariance of general relativity. One point of view, which could be adopted here, is to generalize the framework of Chap. 3, and to allow in (3.5) manifolds M and maps ψ and Ψ, which depend upon the Cauchy data. Such an approach is somewhat too abstract and formal to our taste, so we will instead show here how to formulate the problem in a way which fits the framework of Chap. 3; compare the discussion at the beginning of Sect. 5.3. Thus, the overall set-up here will be that of Chaps. 3 and 4: we start with a model three-dimensional manifold $\widetilde{\Sigma} = \Sigma \cup \Sigma_{\mathscr{I}^+}$, which will be "moved" in a conformally completed manifold $\widetilde{M} = M \cup \mathscr{I}^+$ by various maps $\psi : I \times \widetilde{\Sigma} \to \widetilde{M}$. There is, moreover, a further subtlety related with a choice of *asymptotic* reference frames. As already emphasized in Chap. 3, a Hamiltonian dynamics of fields is always defined *with respect* to a given reference frame, which may be arbitrarily chosen. For scalar fields, considered in Sections 4.4 and 4.5 within the context of special relativity, we have defined the reference frame by the choice of one Lorentzian coordinate system (x^μ) (as done in the preliminary discussion of Sect. 4.1). This choice determined a particular parameterization of $\mathscr{I}^+ \approx \mathrm{R}^1 \times S^2$, whose piece $\Sigma_{\mathscr{I}^+}$ was glued together with a hiperboloid $\mathscr{S}$ in order to construct our abstract initial data surface $\widetilde{\Sigma}$.

In general relativity we will define the asymptotic references frames by choosing a parameterization of an "abstract Scri manifold" $\mathscr{I}^+$, which we are going to glue together with the hyperboloid Σ. Those parameterizations must respect the geometric structure of $\mathscr{I}^+$, which we discuss in Sect. 6.3.

Different parameterizations allowed by this structure differ by the action of a group of transformations, called the BMS group.

We note that we shall not discuss the gravitational analogue of the phase space $\mathscr{P}_{(-\infty,0]}$ considered in Sect. 4.3, as we do not have at our disposal theorems, which would guarantee existence of a sufficiently large class of metrics[11] with properties analogous to the ones considered in Sect. 4.3; in any case such phase spaces have already been discussed in [9, 74, 83].

Let us start by defining a space $\mathscr{P}$, the phase space $\mathscr{P}_{[-1,0]}$ considered in Sect. 5.8 will be a subset thereof; the phase space $\widehat{\mathscr{P}}_{[-1,0]}$ of Sect. 5.9 will actually coincide as a point set with $\mathscr{P}$, but because the dynamics on $\widehat{\mathscr{P}}_{[-1,0]}$ will be different from that on $\mathscr{P}_{[-1,0]} \subset \mathscr{P}$ we prefer to use a different symbol for $\mathscr{P}$ and $\widehat{\mathscr{P}}_{[-1,0]}$. Let Σ be a three-dimensional manifold such that $\bar{\Sigma} = \Sigma \cup \partial\Sigma$ is compact, with $\partial\Sigma \approx S^2$. In a manner somewhat analogous to that of Sects. 4.3–4.5 we let $\mathscr{P}$ be the space of triples $((\tilde{\gamma}, \omega), (\tilde{\pi}, \dot{\omega}), c)$, where[12]

- $\tilde{\gamma} = \tilde{\gamma}_{ij} dx^i \otimes dx^j$ is a Riemannian metric on Σ which can be smoothly extended by continuity to $\partial\bar{\Sigma}$, and ω is a non-negative smooth function on $\bar{\Sigma}$ which vanishes precisely on $\partial\bar{\Sigma}$. The metric $\gamma_{ij} \equiv \omega^{-2}\tilde{\gamma}_{ij}$ at "time" τ can be thought of[13] as the pull-back to Σ of the metric induced on the hypersurfaces $\mathscr{S}_\tau \equiv \psi_\tau(\Sigma) \subset M$, $\psi_\tau = \psi(\tau, \cdot)$, ψ as in (3.5), by the physical space-time metric on M.

[11] Let us, however, mention that the initial data recently constructed by Corvino and Schoen [48, 49] can be used to construct a large class of space-times with a $\mathscr{I}^+$ complete to the past, with $\chi_{AB}(u) = 0$ for all $u \le u_0$ for some u_0. It should then follow from the results in [61, 66] that $\mathscr{I}^+$ will be complete to the future as well if the initial data of [48, 49] are close enough to those for Minkowski space-time. If $\mathscr{P}_{(-\infty,0]}$ is defined as the set of relevant data corresponding to those space-times, with dynamics defined in a way analogous to that in Sect. 4.3 (together with the ingredients described in the current section), then the dynamics will be Hamiltonian, with the ADM mass being the Hamiltonian.

[12] The redundant fields ω and $\dot{\omega}$, which arise as a consequence of the conformal freedom of the framework used here, are only needed to formulate the boundary conditions satisfied by the physically relevant fields (γ_{kl}, P^{kl}) in a simple way.

[13] This interpretation is correct as it stands whenever the map in equation (5.132) below is a diffeomorphism: in such a case the pull-back by this map of the metric on $\widetilde{M}_{((\tilde{\gamma},\omega),(\tilde{\pi},\dot{\omega}),c)}$ defines a metric on M. This isn't true any more at those points of M, at which the map of equation (5.132) is singular, as then the pull-back of the physical metric g degenerates there. In this case the interpretation of γ as the metric induced on the corresponding hypersurface in an appropriate space-time remains correct, but the interpretation of M as a space-time is unjustified. When considering time translations, as we do in this section, there does not seem to be any special reason why one should not restrict oneself to the non-degenerate case. However, for transformations which are tangent to Σ, such as space-translations or rotations, the ζ_τ^{-1}'s will all map Σ to the same hypersurface $\zeta_0^{-1}(\Sigma)$, so that the pull-back of g to M by the map of equation (5.132) will be a tensor field, which has everywhere a signature $(0, +, +, +)$.

- $\tilde{\pi} = \tilde{\pi}^{ij}\partial_i \otimes \partial_j$ is a symmetric tensor density on $\bar{\Sigma}$, which can be smoothly extended by continuity to $\partial \bar{\Sigma}$, while $\dot{\omega}$ is a smooth function on $\bar{\Sigma}$. The physical fields (γ_{kl}, π^{kl}) calculated out of $((\tilde{\gamma}, \omega), (\tilde{\pi}, \dot{\omega}))$ using the formulae of Appendix C.8 (*i.e.*, $\pi^{kl} = \tilde{\pi}^{kl} + 2\sqrt{\det \gamma_{mn}}\gamma^{kl}(\dot{\log \omega}))$ are assumed to satisfy the vacuum constraint equations (5.34)–(5.35).
- For each $u \in [-1, 0]$ the tensor field $c(u) \equiv c_{AB}(u, \cdot)dx^A \otimes dx^B$ is a smooth symmetric tensor field on S^2, which is traceless with respect to the unit round metric $\breve{h}$ on S^2, and varies smoothly in u. The set $[-1, 0] \times S^2$ can be thought of as being a subset of $\mathscr{I}^+$, and then c_{AB} represents the Bondi field χ_{AB} appearing in the expansion (5.98).
- $((\tilde{\gamma}, \omega), (\tilde{\pi}, \dot{\omega}), c)$ satisfy the corner conditions to all orders (*cf.* Appendix C.2), in particular

$$\breve{\lambda}_{AB} = |d\omega|_{\tilde{\gamma}} c_{AB}\Big|_{u=0}\,, \tag{5.129}$$

where $\breve{\lambda}_{AB}$ is the trace-free part of the extrinsic curvature of $\partial\Sigma$, with respect to the inner pointing normal, in the metric $\tilde{\gamma}$. (A finite number of corner conditions would suffice if finite differentiability of the metric tensor on $\widetilde{M}$ is required only.)

As discussed in detail in Appendix C.2, given the data $((\tilde{\gamma}, \omega), (\tilde{\pi}, \dot{\omega}), c)$ described above one can solve the vacuum Einstein equations to obtain a conformally completed space-time $(\widetilde{M}_{((\tilde{\gamma},\omega),(\tilde{\pi},\dot{\omega}),c)}, \tilde{g})$. (We use $((\tilde{\gamma}, \omega), (\tilde{\pi}, \dot{\omega}))$ to construct the relevant Cauchy data on Σ, and use c as the datum for the Bondi tensor field χ on the subset of $[-1, 0] \times S^2 \subset \mathscr{I}^+$, where χ eventually turns out to be defined.) This, then, produces a space-time, which is a development of the data set $((\tilde{\gamma}, \omega), (\tilde{\pi}, \dot{\omega}), c)$. To avoid pathologies it is convenient to require that $(\widetilde{M}_{((\tilde{\gamma},\omega),(\tilde{\pi},\dot{\omega}),c)}, \tilde{g})$ be maximal in the set of globally hyperbolic developments of the Cauchy data; here global hyperbolicity is understood as that of Lorentzian manifolds with boundary.

Now, the known existence theorems for the Cauchy problem at hand do not guarantee that $\mathscr{I}^+$ for the space-time so obtained contains $\mathscr{I}^+_{[-1,0]} \equiv [-1, 0] \times S^2$, where here and throughout the coordinate u in $[-1, 0]$ refers to a Bondi coordinate on $\mathscr{I}^+$ (the restriction to $\mathscr{I}^+$ of the coordinate u of equation (5.93)); one knows only that $\mathscr{I}^+ \supset (-\epsilon, 0] \times S^2$, for some $\epsilon > 0$ but not necessarily $\epsilon > 1$, in general. However, it follows from standard stability results for solutions of hyperbolic partial differential equations that the set of $((\tilde{\gamma}, \omega), (\tilde{\pi}, \dot{\omega}), c)$'s for which

$$\mathscr{I}^+ \supset [-1, 0] \times S^2 \tag{5.130}$$

is open in the set of all such triples, when some natural topology is used on $\mathscr{P}$ (*e.g.*, a Sobolev–type topology). Henceforth we shall assume that some such topology has been chosen, in which existence, uniqueness, and stability of the evolution problem at hand holds. *We shall also assume that (5.130) holds* for all $((\tilde{\gamma},\omega),(\tilde{\pi},\dot{\omega}),c) \in \mathscr{P}$. (We actually expect that (5.130) will always hold under the conditions spelled out above, so that (5.130) does probably not lead to any restrictions anyway, but no rigourous statements of this kind are known to us.)

Let us proceed now to the construction of dynamical systems on $\mathscr{P}$. In this section we will be interested in asymptotic time translations, other dynamical systems can be constructed in a similar way. Let τ_- be any strictly negative constant, and let I be any open interval in $\mathbb{R}$ satisfying $I \supset [\tau_-, 0]$. In order to simplify notation, we shall in what follows assume that $\tau_- = -1$, so that

$$I \supset [-1, 0] , \tag{5.131}$$

however all the subsequent discussion applies with -1 replaced by any strictly negative τ_-. It is convenient to assume that I is bounded, which allows us to apply the known stability properties of solutions of Einstein equations, but there are no fundamental reasons why unbounded I's couldn't occur. Time translations require the introduction of a family of hypersurfaces $\mathscr{S}_\tau \subset \widetilde{M}_{((\tilde{\gamma},\omega),(\tilde{\pi},\dot{\omega}),c)}$, $\tau \in I$, with the following properties:

1. $\partial\mathscr{S}_t$ coincides with $\{u = t\} \times S^2 \subset \mathscr{I}^+$.
2. For each $\tau \in I$ the surface $\mathscr{S}_\tau$ is a spacelike hypersurface in $\widetilde{M}_{((\tilde{\gamma},\omega),(\tilde{\pi},\dot{\omega}),c)}$ uniquely determined by its boundary $\partial\mathscr{S}_t$, and depends in a differentiable way upon the metric.
3. For each $\tau \in I$ there exists a diffeomorphism $\zeta_\tau : \mathscr{S}_\tau \to \Sigma$ such that the map
$$I \times \Sigma \ni (\tau, q) \to \zeta_\tau^{-1}(q) \in \widetilde{M}_{((\tilde{\gamma},\omega),(\tilde{\pi},\dot{\omega}),c)} \tag{5.132}$$
is smooth. (The existence of ζ_0 is guaranteed by the construction of $\widetilde{M}_{((\tilde{\gamma},\omega),(\tilde{\pi},\dot{\omega}),c)}$ described in Appendix C.2).

We emphasize that we do not assume that the $\mathscr{S}_\tau$'s form a foliation. The question of existence of the objects satisfying the above requirements is shortly discussed at the end of this section. We note that different choices of the families $\mathscr{S}_\tau$ will lead to different dynamical systems. Similarly, once the $\mathscr{S}_\tau$'s are chosen, different choices of the ζ_τ's lead to different dynamics. The map considered in point 3 is rather similar to the map ψ of (3.5), except that ψ there takes values in a fixed manifold M, while the map defined in equation (5.132) depends upon the Cauchy data set, with the target manifold depending upon the Cauchy data set.

When asymptotic symmetries other than asymptotic time translations are considered, the condition 1 should be replaced by an obvious analogue thereof, determined by the motion of sections of $\mathscr{I}^+$ associated to the asymptotic symmetry at hand, while conditions 2 and 3 should hold in all cases.

Now, the fact that (5.130) holds does not guarantee yet that (5.131) does for all data sets in $\mathscr{P}$ and, as our final restriction on $\mathscr{P}$, we shall *assume that (5.131) holds*. This leads to an open subset of the space $\mathscr{P}$ previously defined, which will be non-empty if the gauge conditions[14] — that is, the conditions one imposes to single out the hypersurfaces $\mathscr{S}_\tau$ — are chosen in a reasonable way.[15]

As described in detail in Appendix C.2, in $\widetilde{M}_{((\tilde{\gamma},\omega),(\tilde{\pi},\dot{\omega}))}$ we can introduce a six parameter family of Bondi coordinates. In any such coordinate system we can calculate the tensor field $\chi_{AB}(u)$ wherever defined; by construction we have

$$\chi_{AB}(u) = c_{AB}(u) , \qquad u \in [-1, 0]. \tag{5.133}$$

The above ingredients — including a choice of Bondi coordinate system — allow us to construct a dynamical system on "the Σ part of $\mathscr{P}$". As already indicated, in (3.5) we take

$$\widetilde{\Sigma} = \Sigma \cup \Sigma_{\mathscr{I}^+} , \qquad \Sigma_{\mathscr{I}^+} = [-1, 0] \times S^2 . \tag{5.134}$$

We define

$$M \equiv I \times \Sigma , \qquad \widetilde{M} \equiv I \times \widetilde{\Sigma} , \tag{5.135}$$

and we let the map ψ be the identity. We set

$$((\tilde{\gamma}_t, \omega_t), (\tilde{\pi}_t, \dot{\omega}_t)) \equiv ((\tilde{\gamma}, \omega), (\tilde{\pi}, \dot{\omega}))(t, \cdot) , \tag{5.136}$$

where $\omega(t) \equiv \Omega \circ \zeta_t$, $\tilde{\gamma}(t)$ denotes the pull-back by ζ_t^{-1} of the physical metric on $\mathscr{S}_\tau$ to Σ, *etc.*.

It is reasonable to ask how to satisfy all the requirements we have set forth above. A popular candidate for a choice of the family of hypersurfaces $\mathscr{S}_\tau$ arises by requiring that the mean curvature of the $\mathscr{S}_\tau$'s be constant. This condition has the drawback that it restricts the class of allowed $\mathscr{S}_0$'s. It should also be noted that even in smoothly compactifiable space-times such a condition forces one to consider polyhomogeneous initial data sets on Σ, discussed in Sect. 6.10 below, as the solutions of the constant mean curvature equation are expected to be polyhomogeneous at $\mathscr{I}^+$ in general (we emphasize that this does not lead to any difficulties in our approach, *cf.* Sect. 6.10). A related more flexible condition, which one expects to be well posed in some space-time neighbourhood of each initial data hypersurface, is to require that the mean curvature of the $\mathscr{S}_\tau$'s be τ-independent, when followed along the integral curves of the vector of normals to the $\mathscr{S}_\tau$'s; this approach has already

[14] It should be pointed out that (5.130) has a different character as compared to (5.131): (5.130) is a geometric condition because the relevant Bondi coordinates are uniquely defined by the geometry of the problem at hand, while (5.131) is a gauge condition, which depends upon the details of the choices of the hypersurfaces $\mathscr{S}_\tau$.

[15] For "reasonable" gauge choices, for each data set in $\mathscr{P}$ we can find some interval I and some $\tau_- < 0$ such that $I \supset [\tau_-, 0]$. As already mentioned, we can then carry out the construction presented below with $[-1, 0]$ replaced by $[\tau_-, 0]$. In this sense condition (5.131) does not lead to any restriction, and is being made here for convenience of notation only.

been used in Sect. 5.4. The maps ζ_τ can then be constructed by following those integral curves back to the surface $\mathscr{S}_0 \approx \Sigma$. This last device, of following the integral curves of the vector of normals to the $\mathscr{S}_\tau$'s, can always be used whenever the $\mathscr{S}_\tau$'s form a foliation; under the current conditions this should be the case for hypersurfaces obtained by imposing the above discussed restrictions on their extrinsic curvature.

A different approach, which provides a construction of the $\mathscr{S}_\tau$'s and ζ_τ's is to choose some well posed hyperbolic formulation of the conformal Einstein equations, by which we mean a formulation for which existence, uniqueness, and stability of solutions holds. Here "uniqueness" means "absolute uniqueness", as opposed to "uniqueness up to coordinate transformations", thus some gauge choices have to be made, which ensure a unique decomposition of space-time into space and time, together with unique propagation of coordinate systems from one time slice to another. There are quite a few ways of doing that [62–64] — examples of such procedures have been given in the simpler contexts of Sects. 5.3 and 5.4; however, none of those formulations has been studied in detail from the point of view of the evolution problem considered here. Each such system leads in principle to the required unique decomposition of (perhaps a subset of, which is sufficient for our purposes) $\widetilde{M}_{((\tilde{\gamma},\omega),(\tilde{\pi},\dot{\omega}),c)}$ as $I \times \Sigma$, for some open interval I (depending perhaps upon the initial data), which immediately provides the required maps.

5.8 The phase space $\mathscr{P}_{[-1,0]}$

Similarly to Sect. 4.4, we set

$$\mathscr{P}_{[-1,0]} = \Big\{((\tilde{\gamma},\omega),(\tilde{\pi},\dot{\omega}),c) \in \mathscr{P} \quad \text{such that} \\ \frac{\partial c}{\partial u} \quad \text{vanishes in a neighbourhood of} \quad u = -1\Big\} . \tag{5.137}$$

The field c_t is obtained from the equation

$$c_t(u) = \chi(u-t,\cdot) , \qquad u \in [-1,0] , \tag{5.138}$$

analogous to (4.47). Here χ appearing at the right hand side of (5.138) is the Bondi tensor calculated at $\mathscr{I}^+$ in $\widetilde{M}_{((\tilde{\gamma},\omega),(\tilde{\pi},\dot{\omega}),c)}$, and only those t's are allowed, for which χ is defined for all $u \in [-1,0]$; stability of solutions of hyperbolic PDE's guarantees that for each initial data set in $\mathscr{P}_{[-1,0]}$ this holds for t's in an open neighbourhood of $t=0$. We then set

$$T_t((\tilde{\gamma},\omega),(\tilde{\pi},\dot{\omega}),c) \equiv ((\tilde{\gamma}_t,\omega_t),(\tilde{\pi}_t,\dot{\omega}_t),c_t) , \tag{5.139}$$

with $((\tilde{\gamma}_t,\omega_t),(\tilde{\pi}_t,\dot{\omega}_t))$ defined in (5.136). If $\partial c/\partial u$ vanishes in a neighbourhood of $u=-1$, then so will $\partial c_t/\partial u$ for t small enough, and it follows that T_t defines a local dynamical system on $\mathscr{P}_{[-1,0]}$.

In summary, in local coordinates (t, x^i) on M and (u, x^A) on $\mathscr{I}^+$ we have

$$\begin{aligned} \gamma_t(x^i) &= \gamma(t, x^i) , \\ P_t^{ij}(x^i) &= P^{ij}(t, x^i) , \\ c_t(u, x^A) &= \chi(u - t, x^A) . \end{aligned} \tag{5.140}$$

Here $\gamma = \gamma_{ij} dx^i dx^j$ and P^{ij} are the usual ADM fields on Σ.

For fields in $\mathscr{P}_{[-1,0]}$ the boundary integral in (5.128) (with $\tau_- = 1$) involving $\partial\chi/\partial u(u = -1) = \partial c/\partial u(u = -1) = 0$ vanishes, which shows that the local dynamical system $(\mathscr{P}_{[-1,0]}, T_t)$ is Hamiltonian, with

$$H(X, \mathscr{S}) = E_{TB}(\mathscr{S}) = \frac{1}{4\pi} \int_{S^2} M \sin\theta d\theta d\varphi \tag{5.141}$$

— the Trautman–Bondi energy of the hyperboloidal hypersurface $\mathscr{S}$; here X is any vector field, which coincides with ∂_u near $\mathscr{I}^+ \cap \mathscr{S}$. As in the scalar field case there is, however, an *essential ambiguity* which arises if we try to represent H as a functional defined over all field configurations, not only those that have vanishing u-derivatives in a neighbourhood of S_{-1}. Indeed, we can add to H any functional of c and a finite number of its derivatives at S_{-1}, which has the property that it vanishes when $\frac{\partial c}{\partial u}\big|_{S_{-1}} = 0$. From the phase space point of view this will be identical with H, but will not coincide with it for general fields, which are not in $\mathscr{P}_{[-1,0]}$. This is discussed in more detail in Sect. 5.10.

5.9 The phase space $\widehat{\mathscr{P}}_{[-1,0]}$

To obtain the Hamiltonian description of the dynamics in the previous section, we have "killed" the boundary term in (5.128) by imposing the condition $\partial c/\partial u = 0$ in a neighbourhood of $S_{-1} \equiv \{u = -1\}$. It turns out that the Hamiltonian description of the field dynamics on $\mathscr{I}^+$ is also possible without imposing any such conditions, provided one accepts to use *time-dependent* Hamiltonians. Let, as in equation (4.9), $\Sigma_{\mathscr{I}^+} = [-1, 0] \times S^2$ and let

$$\psi_{I \times \Sigma_{\mathscr{I}^+}} : (-1, \infty) \times \Sigma_{\mathscr{I}^+} \to \mathscr{I}^+$$

be any family of maps, which act trivially on the S^2 factor of $\Sigma_{\mathscr{I}^+}$,

$$\psi_{I \times \Sigma_{\mathscr{I}^+}}(\tau, u, x^A) = (\sigma_\tau(u), x^A) \in \mathbb{R} \times S^2 \approx \mathscr{I}^+ .$$

Here $(\sigma_t)_{t > -1}$ is any smooth family of smooth orientation preserving diffeomorphisms from $[-1, 0]$ to $[-1, t]$, with $\sigma_0(u) = u$. Let $\widehat{\mathscr{P}}_{[-1,0]} = \mathscr{P}$, given $t \in I$ and a data set $((\tilde{\gamma}, \omega), (\tilde{\pi}, \dot{\omega}), c) \in \widehat{\mathscr{P}}_{[-1,0]}$ let $\widetilde{M}_{((\tilde{\gamma}_t, \omega_t), (\tilde{\pi}_t, \dot{\omega}_t), c_t)}$ be the space-time defined in Sect. 5.7 using the data $((\tilde{\gamma}_t, \omega_t), (\tilde{\pi}_t, \dot{\omega}_t))$ defined by equation (5.136), while the missing Bondi data χ are provided by c_t defined as

$$c_t(u) = c(\sigma_t(u), \cdot) \, .$$

We read off the Bondi function χ in $\widetilde{M}_{((\tilde{\gamma}_t,\omega_t),(\tilde{\pi}_t,\dot{\omega}_t),c_t)}$ (as already previously described) by introducing Bondi coordinates in a neighbourhood of $\mathscr{I}^+$ in $\widetilde{M}_{((\tilde{\gamma}_t,\omega_t),(\tilde{\pi}_t,\dot{\omega}_t),c_t)}$, and then set

$$c_s(u) = \chi(\sigma_s^{-1}(u), \cdot) \, ,$$

and

$$T_{s,t}(((\tilde{\gamma},\omega),(\tilde{\pi},\dot{\omega}),c)) \equiv ((\tilde{\gamma}_s,\omega_s),(\tilde{\pi}_s,\dot{\omega}_s),c_s) \, . \tag{5.142}$$

As in the scalar field case we have $T_{t,t} = \mathrm{id}$, as expected. A calculation identical to that of Sect. 4.5 leads to the formula

$$\mathscr{H}_{\mathscr{I}^+_{[-1,0]}}(c,\tau) = \frac{1}{32\pi}\int_{-1}^{0} du \int_{S^2} \breve{h}^{AC}\breve{h}^{BD}\frac{\partial c_{AB}}{\partial u}\frac{\partial c_{CD}}{\partial u} Z_\tau(u)\, \sin\theta \, d\theta \, d\varphi \, , \tag{5.143}$$

with $Z_\tau(u)$ as in equation (4.58). Hence

$$H = E_{TB}(\mathscr{S}_\tau) + \mathscr{H}_{\mathscr{I}^+_{[-1,0]}}(c,\tau) \, . \tag{5.144}$$

5.10 Preferred role of the Trautman–Bondi energy

The discussion of the preferred role of the Trautman–Bondi energy is essentially identical to that of Sect. 4.6, so that we will only state the conclusions one can reach in the gravitational field case. In the phase space $\mathscr{P}_{[-1,0]}$ any functional of the form

$$\hat{H}(\mathscr{S}_{-1}) = E_{TB}(\mathscr{S}_{-1}) + \int_{S_{-1}} F\left[\chi|_{S_{-1}}, \frac{\partial\chi}{\partial u}\Big|_{S_{-1}}, \ldots, \frac{\partial^k\chi}{\partial u^k}\Big|_{S_{-1}}\right] + a \tag{5.145}$$

is a Hamiltonian, with a being a constant. We expect (see Appendix C.2) that the fields $\frac{\partial^i\chi}{\partial u^i}\Big|_{S_{-1}}$, $i = 0, \ldots, k$, can take arbitrary values; if that expectation is correct, when we require the function $t \to \hat{H}(\mathscr{S}_t)$ to be monotonic in time t, then the analysis in [43] shows that F has to be constant on $\mathscr{P}_{[-1,0]}$. Hence, under the proviso above, the Trautman–Bondi energy (5.115) is the only Hamiltonian on $\mathscr{P}_{[-1,0]}$, which is monotonic when the hypersurfaces $\mathscr{S}_t$ are moved forward in time in the corresponding space-time.

Considerations similar to those of Sect. 4.6 show that

$$\lim_{\tau\to-1} \mathscr{H}_{\mathscr{I}^+_{[-1,0]}}(c_\tau,\tau) = 0 \, ,$$

hence the limit as τ tends to -1 of the Hamiltonian (5.144) in $\widehat{\mathscr{P}}_{[-1,0]}$ equals the Trautman–Bondi mass (5.115) of the hyperboloidal hypersurface $\mathscr{S}_{-1}$.

6. Hamiltonians associated with the BMS group

6.1 The Poincaré group: convergence of integrals

The aim of this section is to analyse the convergence of the integrals involved for those vector fields X which are Killing vector fields of the background metric. We will consider hypersurfaces $\mathscr{S}$ which, in a neighbourhood $\mathscr{O}$ of $\partial\Sigma$ coordinatized by (x, θ, φ), are of the form

$$\mathscr{S} \cap \mathscr{O} = \{u = \alpha(x, \theta, \varphi)\} , \tag{6.1}$$

with α — a differentiable function of its arguments. Throughout this section we will consider arbitrary vector fields X, which extend continuously and differentiably to $\mathscr{I}^+$, with $x^{-1}X^x$ being $O(1)$. In particular, in several calculations we will not assume that X is a Killing vector of the background unless explicitly stated otherwise.

Let us start with an analysis of the putative Hamiltonian:

$$H(X, \mathscr{S}^\epsilon) = \frac{1}{2}\int_{\partial\mathscr{S}^\epsilon} \mathbb{W}^{\nu\lambda}\, dS_{\nu\lambda} , \tag{6.2}$$

where

$$\mathscr{S}^\epsilon \equiv \mathscr{S} \setminus \{x \leq \epsilon\} . \tag{6.3}$$

On $\partial\mathscr{S}^\epsilon$ we have $du = \alpha_{,\varphi}d\varphi + \alpha_{,\theta}d\theta$, so that $\mathbb{W}^{x\varphi}\, dS_{x\varphi} + \mathbb{W}^{x\theta}\, dS_{x\theta} = -\mathbb{W}^{x\varphi}du \wedge d\theta + \mathbb{W}^{x\theta}du \wedge d\varphi = -\mathbb{W}^{x\varphi}\alpha_{,\varphi}d\varphi \wedge d\theta + \mathbb{W}^{x\theta}\alpha_{,\theta}d\theta \wedge d\varphi = \mathbb{W}^{xA}\partial_A\alpha\, d\theta \wedge d\varphi$, and taking the exterior orientation of $S^2_\epsilon \equiv \partial\mathscr{S}^\epsilon$ into account, one obtains

$$H(X, \mathscr{S}^\epsilon) = \textstyle\int_{S^2_\epsilon}(\mathbb{W}^{xu} - \mathbb{W}^{xA}\partial_A\alpha)\, d\theta\, d\varphi . \tag{6.4}$$

Consider, first, those terms in $\mathbb{W}^{\nu\lambda}$, which contain the background-covariant derivatives of the vector field X (*cf.* equation (5.17)); modulo a multiplicative factor they arise in the combination $g^{\alpha[\nu}\delta^{\lambda]}_\beta X^\beta{}_{;\alpha}$. In some calculations it will be convenient to remove the background from the gravitational field — for this purpose we introduce the following field:

$$\Delta\mathfrak{g}^{\mu\nu} := \mathfrak{g}^{\mu\nu} - \frac{1}{16\pi}\sqrt{|\det b|}\, b^{\mu\nu} . \tag{6.5}$$

We have the trivial equality

$$\sqrt{|\det g|}\, g^{\alpha[\nu}\delta^{\lambda]}_{\beta} X^{\beta}{}_{;\alpha} = \Delta \mathfrak{g}^{\alpha[\nu}\delta^{\lambda]}_{\beta} X^{\beta}{}_{;\alpha} + \sqrt{|\det b|}\, b^{\alpha[\nu}\delta^{\lambda]}_{\beta} X^{\beta}{}_{;\alpha} \,. \tag{6.6}$$

equation (5.11) shows that we have $L|_{g=b} = 0$; similarly we have $\mathbb{W}^{\nu\lambda}{}_{\beta}|_{g=b} = 0$ by (5.18). Further, if X is a Killing vector field of the background, we obtain $H^{\mu}|_{g=b} = 0$ by (5.16). It then follows from (5.16) that for any hypersurface $\hat{\mathscr{S}}$ in the Minkowski space-time and for such X's we have

$$\int_{\partial\hat{\mathscr{S}}} \sqrt{|\det b_{\rho\sigma}|} b^{\alpha[\nu}\delta^{\lambda]}_{\beta} X^{\beta}{}_{;\alpha}\, dS_{\nu\lambda} = 0 \,. \tag{6.7}$$

(An alternative way of obtaining this equation is to notice that this is just Komar's integral for a Killing vector field in Minkowski space-time, which vanishes because Minkowski space-time is Ricci flat.) Now, if we use Bondi coordinates to identify a neighbourhood of $\mathscr{I}^+$ in the space-time into consideration with a neighbourhood of $\mathscr{I}^+$ in Minkowski space-time, then for any $\mathscr{S}^{\epsilon}$ with ϵ small enough there exists a hypersurface $\hat{\mathscr{S}}^{\epsilon}$ in Minkowski space-time such that $\partial\mathscr{S}^{\epsilon} = \partial\hat{\mathscr{S}}^{\epsilon}$. It follows that equation (6.7) with $\hat{\mathscr{S}}$ replaced by $\mathscr{S}^{\epsilon}$ holds for ϵ small enough, in the class of space-times considered here, for X's which are Killing vector fields of the background metric b.

Let us now turn our attention to those terms in the integral (6.4), which arise from $\mathbb{W}^{xu} d\theta d\varphi$; for the convenience of the reader let us repeat here equation (5.114):

$$\begin{aligned} \mathbb{W}^{xu} = \frac{\sin\theta}{16\pi} \Big\{ & X^u \left[x^{-2}\mathscr{D}_A U^A - 2V + x^{-1}\mathrm{e}^{2\beta} h^{AB}\breve{h}_{AB} + \mathrm{e}^{-2\beta} x^{-2} U_A U^A{}_{,x} \right] \\ & - x^{-2} X^u{}_{,A} U^A + (1 - Vx) X^u{}_{,x} + x^{-2} X^x \left[2x^{-1} - x^{-1}\mathrm{e}^{2\beta} h^{AB}\breve{h}_{AB} \right] \\ & - x^{-2} X^A \mathrm{e}^{-2\beta} h_{AB} U^B{}_{,x} + x^{-2}(X^x{}_{;x} - X^u{}_{;u}) - X^u{}_{;x} \Big\} \,. \end{aligned} \tag{6.8}$$

A rough analysis of the behaviour of $\mathbb{W}^{xu}$ can be carried on as follows: first, X^u, $x^{-1}X^x$, are bounded by hypothesis on X. Next, $x^{-2}\beta$, $x^{-2}(h^{AB} - \breve{h}^{AB})\breve{h}_{AB}$, $x^{-2}(\mathrm{e}^{2\beta} - 1)$ and $x^{-2}U^A$ are all $O(1)$ in view of equations (5.98)–(5.100), and this behaviour is preserved under differentiation in the obvious way (recall that (5.98) is an assumption concerning the behaviour of the gravitational field along the hypersurfaces $u = \text{const}$, while (5.99)–(5.100) follow from (5.98) and from the vacuum Einstein equations). This shows convergence — when ϵ tends to zero — of all the integrals arising in (6.4) from $\mathbb{W}^{xu}$, except perhaps for the terms involving the background covariant derivatives of X, and the term $x^{-2}X^A\mathrm{e}^{-2\beta}h_{AB}U^B{}_{,x}$. (Clearly, the latter will vanish if X^A vanishes, which establishes convergence of the appropriate integrals, up to the integrals containing the background covariant derivatives of X, for the "supertranslations" considered in Sect. 6.2). If X^A does not vanish, a more detailed analysis is needed: in the coordinate system $(z^{\mu}) = (u, x^A, x)$ with

$(z^A) = (x^A)$ the Killing vector fields of the background metric (5.96) take the form

$$X_{trans} = -v\frac{\partial}{\partial u} - x^2 v\frac{\partial}{\partial x} + x\breve{h}^{AB}\partial_A v\frac{\partial}{\partial y^B}\,, \tag{6.9}$$

$$X_{rot} = \epsilon^{AB}\partial_B v\frac{\partial}{\partial y^A}\,, \tag{6.10}$$

$$X_{boost} = uX_{trans} - xv\frac{\partial}{\partial x} + \breve{h}^{AB}\partial_A v\frac{\partial}{\partial y^B}\,. \tag{6.11}$$

Here v is a linear combination of $\ell = 1$ spherical harmonics, as in equations (4.69)–(4.71). It can be checked either directly, or using the Killing equations, that for Killing vector fields we have

$$\partial_u X^A = 0 = X_{A\|B} + X_{B\|A} - X^C{}_{\|C}\breve{h}_{AB}\,, \tag{6.12}$$

where $\|$ denotes the covariant derivative on S^2 (previously denoted by $\breve{\mathscr{D}}$ in (5.99)), associated with the round unit metric $\breve{h}_{AB}$. Returning to equation (6.4), suppose first that α is angle-independent; from the explicit form of $x^{-2}U^A\Big|_{\mathscr{I}^+}$ (*cf.* equation (5.99)) and from what has been said, the integral involving $x^{-2}X^A\mathrm{e}^{-2\beta}h_{AB}U^B{}_{,x}$ reads (up to a multiplicative factor)

$$\begin{aligned}
&-2\int_{S^2}\frac{X_A}{x}\chi^{AB}{}_{\|B}\sin\theta\,d\theta\,d\varphi + O(1)\\
&= 2\int_{S^2}\frac{\chi^{AB}}{x}X_{A\|B}\sin\theta\,d\theta\,d\varphi + O(1)\\
&= \int_{S^2}\frac{\chi^{AB}}{x}\left[X_{A\|B} + X_{B\|A} - X^C{}_{\|C}\breve{h}_{AB}\right]\sin\theta\,d\theta\,d\varphi + O(1)\,,
\end{aligned} \tag{6.13}$$

where the symmetry and the vanishing of the $\breve{h}$–trace of χ^{AB} have been used. The vanishing of the dangerous terms follows now from equation (6.12). Further, the terms in the last line of (6.8) are precisely the ones in equation (6.6) which do not involve $\triangle\,\mathfrak{g}$, and equation (6.7) shows that they give a vanishing contribution when integrated upon a sphere. This establishes convergence of $H(X,\mathscr{S}^\epsilon)$ when ϵ tends to zero for vector fields X, which satisfy the requirements spelled out at the beginning of this section, together with (6.12), when $\alpha|_{x=0}$ is a constant, up to the question of convergence of the terms involving the background covariant derivatives of X, for those X's which are not Killing vectors of the background metric. We note that those terms are independent of the physical metric g, and therefore their value is irrelevant from the point of view of the variational formulae considered here; discarding them corresponds to a redefinition of the zero point of the energy.

Using (5.114) and the asymptotics (5.98)–(5.100) of the gravitational field at $\mathscr{I}^+$, in Appendix C.4 we derive the following formula, which holds for all

X's satisfying the asymptotic requirements spelled at the beginning of this section together with (6.12):

$$\int_{S_\tau} \mathbb{W}^{xu}\Big|_{\mathscr{I}^+} d\theta\, d\varphi = \int_{S_\tau} \frac{\sin\theta}{16\pi}\, d\theta\, d\varphi \Big[(4M - \chi^{AB}{}_{\|AB}) X^u - \frac{3}{8} x^{-1} X^x \chi_{AB}\chi^{AB} - \left(6N_A + \frac{1}{2}\chi_{AB}\chi^{BC}{}_{\|C} + \frac{3}{16}(\chi_{CD}\chi^{CD})_{\|A}\right) X^A + x^{-1} X^A \chi_A{}^B{}_{\|B} \Big] + \int_{S_\tau} \mathbb{W}^{xu}|_{g=b}\Big|_{\mathscr{I}^+} d\theta\, d\varphi\,, \qquad (6.14)$$

where the shorthand $\mathbb{W}^{\alpha\beta}\Big|_{\mathscr{I}^+}$ is used for $\lim_{x\to 0}\mathbb{W}^{\alpha\beta}$, similarly for $x^{-1}X^x$; we also have

$$\int_{S_\tau} \mathbb{W}^{xu}|_{g=b}\Big|_{\mathscr{I}^+} d\theta\, d\varphi = \frac{1}{16\pi}\int \left[x^{-2}(X^x{}_{;x} - X^u{}_{;u}) - X^u{}_{;x} \right] \sin\theta\, d\theta\, d\varphi\,. \qquad (6.15)$$

It can be seen[1] that the Trautman–Bondi energy formula (5.115) is a special case of (6.14). More generally, if $X = a$ is a translation of Minkowski space-time which in Minkowskian coordinates x^μ takes the form

$$a = a^\mu \partial_\mu\,,$$

then the corresponding function v in Equation (6.9) reads

$$v(a) = a_0 + a_i n^i\,. \qquad (6.16)$$

Equation (6.9) inserted into (6.14) gives then the following formula for the *Trautman-Bondi four-momentum* [43, 83]:

$$p_{TB}(a, \mathscr{S}) = -\frac{1}{16\pi}\int_{S^2} (4M - \chi^{AB}{}_{\|AB})\, v(a) \sin\theta d\theta d\varphi\,, \qquad (6.17)$$

or, in component notation,

$$p^\mu a_\mu = \frac{1}{16\pi}\int_{S^2} (4M - \chi^{AB}{}_{\|AB})\,(a_0 + a_i n^i) \sin\theta d\theta d\varphi = \frac{1}{4\pi}\int_{S^2} M(a_0 + a_i n^i)\sin\theta d\theta d\varphi\,. \qquad (6.18)$$

To analyse the case with $\alpha|_{x=0}$ — a general differentiable function on S^2, we have the following formula, which is derived in Appendix C.4, *cf.* equation (C.101):

---

[1] The term involving $\mathbb{W}^{xu}|_{g=b}\Big|_{\mathscr{I}^+}$ vanishes for vector fields X associated with the Poincaré group by (6.7); it also vanishes for supertranslations (which follows from (6.15) and (6.31)–(6.32)), hence it vanishes for all the generators of the BMS group. It might, however, give a non-vanishing contribution for more general vector fields.

$$H(X,\mathscr{S}^\epsilon) = \frac{1}{16\pi}\int_{S^2} x^{-1}\left[\chi^{AB}{}_{\|B} + \partial_u\chi^{AB}\partial_B\alpha\right] X_A \sin\theta\, d\theta\, d\varphi + O(1)\,. \tag{6.19}$$

An integration by parts in (6.19), as in equation (6.13), that takes the account that all the functions appearing there are taken at $u = \alpha(\theta,\varphi)$, together with equation (6.12), shows that the potentially divergent integral in (6.19) integrates out to zero, yielding a finite value for $H(X,\mathscr{S}^\epsilon)$.

Let us, next, consider the boundary term of the variational formula (5.21). Equations (5.98)–(5.100) and (C.9)–(C.8) lead to

$$p^x{}_{AB} = \frac{1}{2}x\partial_u\chi_{AB} + O(x^2)\,, \tag{6.20}$$

$$p^A{}_{xB} = -\frac{1}{2}\chi^A{}_B + O(x)\,, \tag{6.21}$$

$$16\pi\delta\mathfrak{g}^{AB} = -x^{-1}\sin\theta\delta\chi^{AB} + O(1)\,, \tag{6.22}$$

$$p^u{}_{AB} = -\chi_{AB} + O(x)\,. \tag{6.23}$$

With this asymptotics the integrand of the boundary integral in (5.21) for a hypersurface $\mathscr{S}^\epsilon$ such that $\partial\mathscr{S}^\epsilon = \{u = \alpha(x,\theta,\varphi), x = \epsilon\}$ takes the form

$$X^u p^x{}_{\mu\nu}\delta\mathfrak{g}^{\mu\nu} - X^x p^u{}_{\mu\nu}\delta\mathfrak{g}^{\mu\nu} + \partial_A\alpha\left(X^x p^A{}_{\mu\nu}\delta\mathfrak{g}^{\mu\nu} - X^A p^x{}_{\mu\nu}\delta\mathfrak{g}^{\mu\nu}\right) =$$

$$-\frac{\sin\theta}{16\pi}\left[x^{-1}X^x\chi_{AB} + (X^u - X^C\partial_C\alpha)\frac{1}{2}\partial_u\chi_{AB}\right]\delta\chi^{AB} + O(x)\,. \tag{6.24}$$

It follows again that, under the current asymptotic conditions, the boundary integral in (5.21) converges when ϵ tends to zero. (We note that the fact that X is a Killing vector of the background has not been used in (6.20)–(6.24).)

Let us, finally, pass to the volume integral in (5.21). In a neighbourhood $\mathscr{O}$ of $\partial\Sigma$ coordinatized by (x,θ,φ) we have

$$\begin{aligned}
&\int_{\mathscr{S}\cap\mathscr{O}} \left(\pounds_X p^\lambda{}_{\mu\nu}\delta\mathfrak{g}^{\mu\nu} - \pounds_X\mathfrak{g}^{\mu\nu}\delta p^\lambda{}_{\mu\nu}\right) dS_\lambda = \\
&\int_{\mathscr{S}\cap\mathscr{O}} \Big(\pounds_X p^u{}_{\mu\nu}\delta\mathfrak{g}^{\mu\nu} - \pounds_X\mathfrak{g}^{\mu\nu}\delta p^u{}_{\mu\nu} \\
&-\partial_k\alpha(\pounds_X p^k{}_{\mu\nu}\delta\mathfrak{g}^{\mu\nu} - \pounds_X\mathfrak{g}^{\mu\nu}\delta p^k{}_{\mu\nu})\Big)\, dx\, d\theta\, d\varphi\,.
\end{aligned}$$

We have $\delta\mathfrak{g} = \delta\,\Delta\mathfrak{g}$; further, if X is a Killing vector field for the background metric, then $\pounds_X\mathfrak{g} = \pounds_X\,\Delta\mathfrak{g}$. This allows us to rewrite the last volume integral in the form

$$\begin{aligned}
&\int_{\mathscr{S}\cap\mathscr{O}} \left(\pounds_X p^\lambda{}_{\mu\nu}\delta\mathfrak{g}^{\mu\nu} - \pounds_X\mathfrak{g}^{\mu\nu}\delta p^\lambda{}_{\mu\nu}\right) dS_\lambda = \\
&\int_{\mathscr{S}\cap\mathscr{O}} \Big(\pounds_X p^u{}_{\mu\nu}\delta\,\Delta\mathfrak{g}^{\mu\nu} - \pounds_X\,\Delta\mathfrak{g}^{\mu\nu}\delta p^u{}_{\mu\nu} \\
&-\partial_k\alpha(\pounds_X p^k{}_{\mu\nu}\delta\,\Delta\mathfrak{g}^{\mu\nu} - \pounds_X\,\Delta\mathfrak{g}^{\mu\nu}\delta p^k{}_{\mu\nu})\Big)\, dx\,.
\end{aligned} \tag{6.25}$$

In what follows we shall analyse the convergence of the right hand side of the equation above for the class of vector fields X as spelled out at the beginning of this section, not necessarily Killing vector fields of the background. We have the following asymptotics of the $\Delta\mathfrak{g}^{\mu\nu}$'s:

$$16\pi\delta\ \Delta\mathfrak{g}^{AB} = -x^{-1}\sin\theta\delta\chi^{AB} + O(1)\,,$$

$$\Delta\mathfrak{g}^{xx} = O(x)\,,\quad \Delta\mathfrak{g}^{xA} = O(1)\,,\quad \Delta\mathfrak{g}^{u\mu} = 0\,.$$

All the $p^\lambda{}_{\mu\nu}$ not listed in (6.20)–(6.23) are $O(x)$. The Lie derivatives of the momenta $p^\lambda{}_{\mu\nu}$,

$$\pounds_X p^\lambda{}_{\mu\nu} = X^\sigma\partial_\sigma p^\lambda{}_{\mu\nu} + p^\lambda{}_{\sigma\nu}\partial_\mu X^\sigma + p^\lambda{}_{\mu\sigma}\partial_\nu X^\sigma - p^\sigma{}_{\mu\nu}\partial_\sigma X^\lambda\,,$$

behave as $O(1)$ or better, while those of the $\Delta\mathfrak{g}^{\mu\nu}$'s,

$$\pounds_X\ \Delta\mathfrak{g}^{\mu\nu} = \partial_\lambda(X^\lambda\ \Delta\mathfrak{g}^{\mu\nu}) - \Delta\mathfrak{g}^{\lambda\nu}\partial_\lambda X^\mu - \Delta\mathfrak{g}^{\mu\lambda}\partial_\lambda X^\nu\,,$$

behave typically as $O(x^{-1})$. In detail we have

$$\pounds_X\ \Delta\mathfrak{g}^{AB} = \partial_u(X^u\ \Delta\mathfrak{g}^{AB}) + \overset{(2)}{\pounds_X}\ \Delta\mathfrak{g}^{AB} + O(1)\,,$$

$$\pounds_X\ \Delta\mathfrak{g}^{uA} = \Delta\mathfrak{g}^{AB}\partial_B X^u + O(1)\,,$$

$$\pounds_X\ \Delta\mathfrak{g}^{xx} = O(x)\,,\quad \pounds_X\ \Delta\mathfrak{g}^{ux} = O(1) = \pounds_X\ \Delta\mathfrak{g}^{xA}\,,\quad \pounds_X\ \Delta\mathfrak{g}^{uu} = 0\,.$$

Here $\overset{(2)}{\pounds_X}$ denotes the Lie derivative on the sphere taken with respect to $X^A\partial_A$ — the "sphere part" of the vector X. Since $\delta p^\lambda{}_{uA} = O(x)$, it follows that the dangerous term in (6.25) is of the form

$$\pounds_X p^u{}_{AB}\delta\ \Delta\mathfrak{g}^{AB} - \pounds_X\ \Delta\mathfrak{g}^{AB}\delta p^u{}_{AB}\,.$$

We further have

$$\pounds_X p^u{}_{AB} = X^u\partial_u p^u{}_{AB} - p^u{}_{AB}\partial_u X^u + \overset{(2)}{\pounds_X} p^u{}_{AB} + O(x)\,.$$

Inserting all this into (6.25), one finds that

$$\begin{aligned}
&\int_{\mathscr{S}\cap\mathscr{O}}\left(\pounds_X p^\lambda{}_{\mu\nu}\delta\mathfrak{g}^{\mu\nu} - \pounds_X\mathfrak{g}^{\mu\nu}\delta p^\lambda{}_{\mu\nu}\right) dS_\lambda\\
&= O(1) - \int_{\mathscr{S}\cap\mathscr{O}}\left\{\left[\partial_u(X^u\ \Delta\mathfrak{g}^{AB}) + \overset{(2)}{\pounds_X}\ \Delta\mathfrak{g}^{AB}\right]\delta p^u{}_{AB}\right.\\
&\left. - \left[X^u\partial_u p^u{}_{AB} - p^u{}_{AB}\partial_u X^u + \overset{(2)}{\pounds_X} p^u{}_{AB}\right]\delta\ \Delta\mathfrak{g}^{AB}\right\} dx\, d\theta\, d\varphi\,. \quad (6.26)
\end{aligned}$$

The terms in the curly bracket above are potentially of order $O(x^{-1})$, more precisely

$$16\pi x\{\ldots\} = O(x) + \left[\partial_u(X^u \sin\theta\chi^{AB}) + \overset{(2)}{\mathcal{L}_X}(\sin\theta\chi^{AB})\right]\delta\chi_{AB}$$

$$-\left[X^u\partial_u\chi_{AB} - \chi_{AB}\partial_u X^u + \overset{(2)}{\mathcal{L}_X}\chi_{AB}\right]\delta(\chi^{AB}\sin\theta)$$

$$= O(x) + \left[2\partial_u X^u \sin\theta\chi^{AB} + \chi_{CD}\overset{(2)}{\mathcal{L}_X}\left(\sin\theta\breve{h}^{AC}\breve{h}^{BD}\right)\right]\delta\chi_{AB}\,. \qquad (6.27)$$

It turns out that the terms in the square bracket in this last equation vanish for Killing vectors: indeed, since $\sin\theta = \sqrt{\det\breve{h}_{AB}}$, we can write

$$2\partial_u X^u \sin\theta\chi^{AB} + \chi_{CD}\overset{(2)}{\mathcal{L}_X}\left(\sin\theta\breve{h}^{AC}\breve{h}^{BD}\right)$$

$$= \left(2\partial_u X^u + X^C{}_{\|C}\right)\sin\theta\chi^{AB} + 2\sin\theta\,\overset{(2)}{\mathcal{L}_X}\breve{h}^{C(A}\chi^{B)}{}_C\,.$$

If X is a rotation, we have $X^u = X^C{}_{\|C} = \overset{(2)}{\mathcal{L}_X}\breve{h}^{AB} = 0$; for translations and supertranslations $\partial_u X^u = 0 = X^A$. This implies our claim for such Killing vectors. In the case of boosts along the z axis $\partial_u X^u = -\cos\theta$, $X^A = \breve{h}^{AB}\partial_A\cos\theta$, and the formula $\overset{(2)}{\mathcal{L}_X}\breve{h}^{CD} = 2\cos\theta\breve{h}^{CD}$ (*cf.* equation (6.12)) guarantees the vanishing of the square bracket in (6.27). We have thus shown that the volume integral in (5.21) converges as well when ϵ tends to zero.

6.2 Supertranslations (and space translations): convergence of integrals

In this section we will check the convergence of the integrals related to the "supertranslations"; recall that translations are understood as a special case of "supertranslations". It will then follow that the local dynamical system generated by supertranslations is Hamiltonian on appropriate phase spaces. The "supertranslation" vector fields are defined by the requirements that: 1) they extend smoothly to $\mathscr{I}^+$; 2) they preserve the Bondi form of the metric; and 3) $X|_{\mathscr{I}^+} = \lambda(\theta,\varphi)\partial_u$ for some smooth function λ on S^2. It may be proved that in the coordinate system of (5.93) such a field takes the following form:

$$X_{supertr} = [\lambda + O(x)]\frac{\partial}{\partial u} - \frac{1}{2}x^2[\Delta_2\lambda + O(x)]\frac{\partial}{\partial x}$$
$$-x[\breve{h}^{AB}\partial_A\lambda + O(x)]\frac{\partial}{\partial x^B}\,, \qquad (6.28)$$

where $\lambda = \lambda(x^A)$ is a smooth function of its arguments, while Δ_2 is the Laplacian of the standard round metric $\breve{h}_{AB}$ on a two-dimensional unit sphere. We shall perform a detailed convergence analysis for the vector field X defined as

$$X = \lambda\frac{\partial}{\partial u} - \frac{1}{2}x^2\Delta_2\lambda\frac{\partial}{\partial x} - x\breve{h}^{AB}\partial_A\lambda\frac{\partial}{\partial x^B}\,, \tag{6.29}$$

which is an exact formula for a supertranslation in Minkowski space-time (see (4.79)); it is easily seen that the error terms neglected lead to convergent integrals. Replacing r by $1/x$ in the flat metric (5.96), we have

$$b = b_{\mu\nu}\,dx^\mu\,dx^\nu = -\,du^2 + 2x^{-2}\,du\,dx + x^{-2}\breve{h}_{AB}\,dx^A\,dx^B\,, \tag{6.30}$$

so that $\sqrt{|\det b|} = x^{-4}\sin\theta$, and of course[8] $B^\sigma{}_{\beta\gamma\kappa} = 0$. The $B^\sigma{}_{\beta\gamma}$'s can be found in Appendix C.1, equations (C.2)–(C.3).

Covariant derivatives $X^\mu{}_{;\nu} = X^\mu{}_{,\nu} + B^\mu{}_{\nu\lambda}X^\lambda$ of fields given by (6.29) read

$$0 = X^0{}_{;\mu} = X^3{}_{;3} = X^3{}_{;0} = X^A{}_{;0} = X^A{}_{;3} = X^\mu{}_{;\mu}\,, \tag{6.31}$$

$$X^3{}_{;A} = -\frac{1}{2}x^2\partial_A(\Delta_2+2)\lambda, \quad X^A{}_{;B} = \frac{1}{2}x(2\lambda^{\|A}{}_B - \delta^A_B\Delta_2\lambda)\,. \tag{6.32}$$

We note that several calculations of Sect. 6.1 go through for supertranslations (6.28) or (6.29), since they satisfy the general requirements spelled out at the beginning of that section. However, because they are not Killing vector fields of the background metric (5.96), the formula for the Hamiltonian (5.15) as well as the variational formula (5.21) will contain supplementary terms, which we will analyse here. Now, convergence of the boundary integral in (5.21) (*cf.* equation (6.24)) follows already from the analysis in Sect. 6.1. Next, consider the g independent terms (arising from the decomposition (6.6)) in the integral H_{boundary} defined by equation (5.25); from (6.31)–(6.32) one finds

$$\begin{aligned}\lim_{\epsilon\to0} H_{\text{boundary}}(X,\mathscr{S}^\epsilon)\Big|_{g=b} &= \lim_{\epsilon\to0}\int_{S^2_\epsilon}\left(\mathbb{W}^{xu} - \alpha_{,A}\mathbb{W}^{xA}\right)\Big|_{g=b}d\theta\,d\varphi\\ &= -\frac{1}{32\pi}\int_{S^2}\left(\lambda\Delta_2(\Delta_2+2)\alpha(0,x^A)\right)\sin\theta\,d\theta\,d\varphi\,. \end{aligned}\tag{6.33}$$

Here and throughout this section, $\mathscr{S}^\epsilon$ is given by equation (6.3), with $\mathscr{S}$ as in equation (6.1). We note that this integral vanishes, when λ or α are a linear combination of $\ell = 0$ or $\ell = 1$ harmonics, consistently with our previous proof that this integral vanishes for X's which are Killing vector fields of the background metric. While it is not necessary to do that, it seems reasonable to normalize the Hamiltonians so that they vanish on the Minkowski background, therefore we set

$$H^{\text{normalised}}_{\text{boundary}}(X,\mathscr{S}) \equiv H_{\text{boundary}}(X,\mathscr{S}) - H_{\text{boundary}}(X,\mathscr{S})\Big|_{g=b}$$

$$= H_{\text{boundary}}(X, \mathscr{S}) + \frac{1}{8\pi}\int_{\partial\mathscr{S}} \sqrt{|\det b_{\rho\sigma}|}b^{\alpha[\nu}\delta^{\lambda]}_{\beta} X^{\beta}{}_{;\alpha}\, dS_{\nu\lambda}$$

$$= \int_{\partial\mathscr{S}} \left[\mathbb{W}^{\nu\lambda}{}_{\beta}X^{\beta} - 2\,\Delta\mathfrak{g}^{\alpha[\nu}\delta^{\lambda]}_{\beta}X^{\beta}{}_{;\alpha}\right] dS_{\nu\lambda}$$

$$= \int_{\partial\mathscr{S}} \Bigg[\left(2\mathfrak{g}^{\mu[\nu}p^{\lambda]}_{\mu\beta} - 2\delta^{[\nu}_{\beta}p^{\lambda]}_{\mu\sigma}\mathfrak{g}^{\mu\sigma} - \frac{2}{3}\mathfrak{g}^{\mu[\nu}\delta^{\lambda]}_{\beta}p^{\sigma}_{\mu\sigma}\right)X^{\beta}$$

$$- 2\,\Delta\mathfrak{g}^{\alpha[\nu}\delta^{\lambda]}_{\beta}X^{\beta}{}_{;\alpha}\Bigg] dS_{\nu\lambda}\,, \tag{6.34}$$

with $\Delta\,\mathfrak{g}$ defined by (6.5). The convergence of $H^{\text{normalised}}_{\text{boundary}}(X, \mathscr{S}^{\epsilon})$ when ϵ tends to 0, for hypersurfaces of the form (6.1), follows from equation (C.101) because $x^{-1}X^{A} = O(1)$.

It remains to analyse what happens, when ϵ tends to zero, with the volume part of the Hamiltonian (see equations (5.24)–(5.26)), and with the volume integral

$$\int_{\mathscr{S}^{\epsilon}} (\pounds_{X}p^{\lambda}{}_{\mu\nu}\delta\mathfrak{g}^{\mu\nu} - \pounds_{X}\mathfrak{g}^{\mu\nu}\delta p^{\lambda}{}_{\mu\nu})\, dS_{\lambda}\,. \tag{6.35}$$

Here, as before, $\mathscr{S}^{\epsilon}$ is defined by equation (6.3). It will be seen that the volume term in the Hamiltonian,

$$H_{\text{volume}}(X, \mathscr{S}^{\epsilon}) := -\int_{\mathscr{S}^{\epsilon}} 2\mathfrak{g}^{\beta[\gamma}\delta^{\mu]}_{\sigma}(X^{\sigma}{}_{;\beta\gamma} - B^{\sigma}{}_{\beta\gamma\kappa}X^{\kappa})\, dS_{\mu}\,,$$

diverges as ϵ goes to zero — this is not surprising in view of the fact that this is also the case with the volume integral (6.35). We will cure this problem by a suitable redefinition of variables, and a corresponding change of Hamiltonian. Let $\Delta\,\mathfrak{g}^{\mu\nu}$ be defined by equation (6.5); using this variable, we can rewrite (5.21) as

$$-\delta H = \int_{\mathscr{S}^{\epsilon}} (-\pounds_{X}\mathfrak{g}^{\mu\nu}\delta p^{\lambda}{}_{\mu\nu} + \pounds_{X}p^{\lambda}{}_{\mu\nu}\delta\mathfrak{g}^{\mu\nu})\, dS_{\lambda}$$

$$+ \int_{\partial\mathscr{S}^{\epsilon}} X^{[\mu}p^{\nu]}{}_{\alpha\beta}\delta\mathfrak{g}^{\alpha\beta}\, dS_{\mu\nu}$$

$$= \int_{\mathscr{S}^{\epsilon}} (-\pounds_{X}\,\Delta\mathfrak{g}^{\mu\nu}\delta p^{\lambda}{}_{\mu\nu} + \pounds_{X}p^{\lambda}{}_{\mu\nu}\delta\,\Delta\mathfrak{g}^{\mu\nu})\, dS_{\lambda}$$

$$-\delta F + \int_{\partial\mathscr{S}^{\epsilon}} X^{[\mu}p^{\nu]}{}_{\alpha\beta}\delta\mathfrak{g}^{\alpha\beta}\, dS_{\mu\nu}\,, \tag{6.36}$$

where

$$F(X, \mathscr{S}^{\epsilon}) := \frac{1}{16\pi}\int_{\mathscr{S}^{\epsilon}} \pounds_{X}\left(\sqrt{|\det b|}b^{\mu\nu}\right)p^{\lambda}{}_{\mu\nu}\, dS_{\lambda}\,.$$

In the notation of equations (5.25)–(5.26) this last equation can be rewritten as

$$\delta(H^{\text{normalised}}_{\text{boundary}} + H_{\text{volume}} - F)(X, \mathscr{S}^{\epsilon}) =$$

$$\int_{\mathscr{S}^\epsilon} \left(\pounds_X \, \Delta\mathfrak{g}^{\mu\nu}\delta p^\lambda{}_{\mu\nu} - \pounds_X p^\lambda{}_{\mu\nu}\delta \, \Delta\mathfrak{g}^{\mu\nu}\right) dS_\lambda - \int_{\partial\mathscr{S}^\epsilon} X^{[\mu}p^{\nu]}{}_{\alpha\beta}\delta\mathfrak{g}^{\alpha\beta}\, dS_{\mu\nu} \,, \tag{6.37}$$

The right hand side of this equation converges as ϵ goes to zero by equations (6.24), (6.26) and (6.27). It turns out that the left hand side is still divergent, but the divergence can be gotten rid of by a "renormalisation" procedure. In order to show that, we note the identity

$$\pounds_X \left(\sqrt{|\det b|}b^{\mu\nu}\right) = \sqrt{|\det b|}\left(b^{\mu\nu}X^\lambda{}_{;\lambda} - 2X^{(\mu;\nu)}\right) ,$$

hence

$$F = \frac{1}{16\pi}\int_{\mathscr{S}^\epsilon} \sqrt{|\det b|}\left(b^{\mu\nu}X^\alpha{}_{;\alpha} - 2X^{(\mu;\nu)}\right) p^\lambda{}_{\mu\nu}\, dS_\lambda \,.$$

On the other hand

$$H_{\mathrm{volume}} = -\int_{\mathscr{S}^\epsilon} \left(\mathfrak{g}^{\beta\gamma}X^\mu{}_{;\beta\gamma} - \mathfrak{g}^{\beta\mu}X^\gamma{}_{;\beta\gamma}\right) dS_\mu + O(1) \,,$$

where the $O(1)$ terms arise from the background curvature which, by hypothesis, is compactly supported. Now, generically we have the behavior $p^\lambda{}_{\mu\nu} = O(1)$, $\mathfrak{g}^{\mu\nu} = O(x^{-2})$, $\Delta\,\mathfrak{g}^{\mu\nu} = O(x^{-1})$, $\sqrt{|\det b|}b^{\mu\nu} = O(x^{-2})$, so that if X is such that $X^\mu{}_{;\nu} = O(x)$ (which is the case for supertranslations), then $H_{\mathrm{volume}}(X,\mathscr{S}^\epsilon)$ and $F(X,\mathscr{S}^\epsilon)$ will typically diverge when ϵ goes to zero: From Equations (6.31)–(6.32) and from $p^u_{AB} = O(1)$, $p^C_{AB} = O(x)$ one finds that the only possibly divergent contribution to F comes from $p^u{}_{AB}$. Equation (6.23), together with the formula

$$\sqrt{|\det b|}X^{A;B} = x^{-1}\sin\theta\left(\lambda^{||AB} - \frac{1}{2}\breve{h}^{AB}\Delta_2\lambda\right)$$

(*cf.* (6.32)), shows that

$$F = \frac{1}{32\pi}\int_{\mathscr{S}^\epsilon} x^{-1}\sin\theta\left(2\lambda^{||AB} - \breve{h}^{AB}\Delta_2\lambda\right)\chi_{AB}\, dx\, d\theta\, d\varphi + O(1) \,, \tag{6.38}$$

where the $O(1)$ terms are finite in the limit $\epsilon \to 0$. Next, it is useful to rewrite H_{volume} as

$$\begin{aligned} H_{\mathrm{volume}}(X,\mathscr{S}^\epsilon) &= -\int_{\mathscr{S}^\epsilon} \left(\Delta\mathfrak{g}^{\beta\gamma}X^\mu{}_{;\beta\gamma} - \Delta\mathfrak{g}^{\beta\mu}X^\kappa{}_{;\beta\kappa}\right) dS_\mu \\ &\quad -\frac{1}{16\pi}\int_{\mathscr{S}^\epsilon} \sqrt{|\det b|}\left(b^{\beta\gamma}X^\mu{}_{;\beta\gamma} - b^{\beta\mu}X^\kappa{}_{;\beta\kappa}\right) dS_\mu + O(1) \,, \end{aligned} \tag{6.39}$$

where $O(1)$ corresponds to background curvature terms, which are irrelevant for the discussion here. To analyse the second integral above, we note the identities

$$\sqrt{|\det b|}\, b^{\beta\gamma}X^A{}_{;\beta\gamma} = x^{-1}\sin\theta(\Delta_2 + 1)\lambda^{||A} \,,$$

$$\sqrt{|\det b|}\, b^{\beta\gamma} X^3{}_{;\beta\gamma} = -\frac{1}{2}\sin\theta \Delta_2(\Delta_2+2)\lambda\,,$$

$$\sqrt{|\det b|}\, b^{\beta\gamma} X^0{}_{;\beta\gamma} = 0; \quad X^\kappa{}_{;\beta\kappa} = 0\,,$$

$$\int_{\mathscr{S}^\epsilon} \sqrt{|\det b|}\, b^{\beta\gamma} X^\mu{}_{;\beta\gamma}\, dS_\mu = \int_{\mathscr{S}^\epsilon} x^{-1}\sin\theta \alpha_{,A}(\Delta_2+1)\lambda^{\|A}\, dx\, d\theta\, d\varphi + O(1)\,.$$

It follows that the second integral in (6.39) diverges when ϵ tends to zero. However, this is a rather harmless divergence from a Hamiltonian point of view, since this integral is independent of the dynamical variables $\mathfrak{g}^{\beta\gamma}$ and $p^\lambda{}_{\mu\nu}$. The corresponding divergence corresponds to a drastically bad choice of the zero point of energy, and is gotten rid of by subtracting the second integral from the formula for the Hamiltonian on $\mathscr{S}^\epsilon$.

Let us turn our attention now to the first integral in (6.39):

$$\begin{aligned} H^{\text{normalised}}_{\text{volume}}(X,\mathscr{S}^\epsilon) &:= -\int_{\mathscr{S}^\epsilon} \left(\Delta\mathfrak{g}^{\beta\gamma} X^\mu{}_{;\beta\gamma} - \Delta\mathfrak{g}^{\beta\mu} X^\kappa{}_{;\beta\kappa}\right) dS_\mu \\ &= -\int_{\mathscr{S}^\epsilon} \Delta\mathfrak{g}^{AB} X^\mu{}_{;AB}\, dS_\mu + O(1)\,. \end{aligned}$$

The dangerous terms above,

$$X^\mu{}_{;AB} = (X^\mu{}_{;A})_{,B} + B^\mu{}_{\lambda B} X^\lambda{}_{;A} - B^\lambda{}_{AB} X^\mu{}_{;\lambda}\,,$$

have the following components:

$$X^0{}_{;AB} = \lambda_{\|AB} - \frac{1}{2}\breve{h}_{AB}\Delta_2\lambda\,,$$

$$X^3{}_{;AB} = -\frac{1}{2}x^2\left[(\Delta_2\lambda)_{\|AB} + \breve{h}_{AB}\Delta_2\lambda\right]\,,$$

$$X^C{}_{;AB} = x\left[\lambda^{\|C}{}_{AB} + \delta^C_B\lambda_{\|A} + (\Delta_2\lambda)_{\|[A}\delta^C_{B]}\right]\,.$$

It follows that

$$\Delta\mathfrak{g}^{AB} X^0{}_{;AB} = \frac{\sin\theta}{16\pi x}\chi^{AB}\left(\frac{1}{2}\breve{h}_{AB}\Delta_2\lambda - \lambda_{\|AB}\right) + O(1)\,,$$

and

$$H^{\text{normalised}}_{\text{volume}}(X,\mathscr{S}^\epsilon) = \int_{\mathscr{S}^\epsilon} \frac{\sin\theta}{16\pi x}\chi^{AB}\left(\lambda_{\|AB} - \frac{1}{2}\breve{h}_{AB}\Delta_2\lambda\right) + O(1)\,.$$

equation (6.38) gives

$$H^{\text{normalised}}_{\text{volume}}(X,\mathscr{S}^\epsilon) - F(X,\mathscr{S}^\epsilon) = O(1)\,.$$

In conclusion, we have a well defined variational formula

$$-\delta\tilde{H}(X,\mathscr{S}) = \int_{\mathscr{S}} \left(\pounds_X p^\lambda{}_{\mu\nu}\delta\,\Delta\mathfrak{g}^{\mu\nu} - \pounds_X\,\Delta\mathfrak{g}^{\mu\nu}\delta p^\lambda{}_{\mu\nu}\right) dS_\lambda$$
$$+ \int_{\partial\mathscr{S}} X^{[\mu}p^{\nu]}{}_{\alpha\beta}\delta\,\Delta\mathfrak{g}^{\alpha\beta}\, dS_{\mu\nu}\,, \tag{6.40}$$

with

$$\tilde{H}(X,\mathscr{S}) \equiv \lim_{\epsilon\to 0}(H^{\text{normalised}}_{\text{boundary}} + H^{\text{normalised}}_{\text{volume}} - F)(X,\mathscr{S}^\epsilon) \tag{6.41}$$

$$= \int_{\mathscr{S}} \left\{\Delta\mathfrak{g}^{\beta\lambda}X^\kappa{}_{;\beta\kappa} - \Delta\mathfrak{g}^{\beta\gamma}X^\lambda{}_{;\beta\gamma} - \frac{\sqrt{|\det b|}}{16\pi}\left(b^{\mu\nu}X^\kappa{}_{;\kappa} - 2X^{(\mu;\nu)}\right)p^\lambda{}_{\mu\nu}\right\} dS_\lambda$$
$$+ \int_{\partial\mathscr{S}} \left[\left(2\mathfrak{g}^{\mu[\nu}p^{\lambda]}_{\mu\beta} - 2\delta^{[\nu}_\beta p^{\lambda]}_{\mu\sigma}\mathfrak{g}^{\mu\sigma} - \frac{2}{3}\mathfrak{g}^{\mu[\nu}\delta^{\lambda]}_\beta p^\sigma_{\mu\sigma}\right)X^\beta\right.$$
$$\left. - 2\,\Delta\mathfrak{g}^{\alpha[\nu}\delta^{\lambda]}_\beta X^\beta{}_{;\alpha}\right] dS_{\nu\lambda}\,. \tag{6.42}$$

In particular, it follows that supertranslations generate a Hamiltonian flow on phase spaces constructed in a manner analogous to that for energy.

6.3 The abstract Scri

So far we have introduced $\mathscr{I}^+$ as the set $x=0$ in a Bondi coordinate system (u,r,θ,φ), with $x = 1/r$. The Bondi coordinate systems differ from each other by Bondi-Metzner-Sachs (BMS) coordinate transformations: the Lorentz transformations, together with the translations and the supertranslations, discussed in Sect. 6.2.[2] As a manifold $\mathscr{I}^+$ is just $\mathbb{R}\times S^2$; in the present section we want to equip $\mathbb{R}\times S^2$ with an abstract structure so that the BMS group arises in a natural way as its symmetry group, *i.e.*, a group of transformations preserving this structure (compare [67, 106, 111]).

First, in the conformally completed manifold with metric x^2g, $\mathscr{I}^+$ is a collection of future-oriented, null geodesics — "light rays at infinity". Each Bondi coordinate u, together with the future-oriented vector $\partial/\partial u$, when restricted to $\mathscr{I}^+$, can be used to define an affine structure and orientation on each of those geodesics. Thus, every Bondi coordinate system endows $\mathscr{I}^+$ with the structure of an affine fiber bundle $\pi:\mathscr{I}^+\to B$, where each fiber $\pi^{-1}(b)$, $b\in B$, is a one-dimensional, oriented affine manifold, with the base manifold B being diffeomorphic to S^2 ("sphere at infinity"). The BMS

[2] The collection of coordinate transformations which preserve the Bondi-Sachs form of the metric is sometimes called *the BMS group*, even though of course there is no group structure involved because the corresponding coordinate systems are only defined locally, they cover different domains, and cannot therefore be composed with each other in general. However, it makes sense to talk about the BMS group as the remnant on $\mathbb{R}\times S^2$ of those coordinate transformations.

transformations preserve this fiber structure, as well as the affine structure and orientation of each fiber.

Now, the base manifold B is not equipped with a metric structure, because the boost transformations of $\mathscr{I}^+$, generated by the restrictions to $\mathscr{I}^+$ of the vector fields (6.11), do not preserve the metric $\breve{h}_{AB}$, but lead to a conformal rescaling. Another way of seeing that is that the tensor field induced on $\mathscr{I}^+$ by the metric $x^2 g$ is rescaled by a conformal factor under change of the Bondi coordinate system. However, these rescalings do not change the conformal structure induced on B by the Bondi system. We conclude that B is equipped with a well defined conformal structure, which we will denote by $\mathscr{C}$.

Consider two Bondi coordinate systems which differ by a boost transformation, the transition from one coordinate system to the other determines a conformal transformation F from S^2 to itself. Treating the sphere as embedded in $\mathbb{R}^3$,

$$S^2 = \left\{ z \in \mathbb{R}^3 \mid (z^1)^2 + (z^2)^2 + (z^3)^2 = 1 \right\} ,$$

every boost transformation $z \to \bar{z} = F(z)$ may be written as follows:

$$\bar{z}^i = \frac{\sqrt{1 - \|v\|^2} z^i - v^i + \frac{v^i v_k z^k}{1 + \sqrt{1 - \|v\|^2}}}{1 - v_k z^k} , \tag{6.43}$$

for a certain vector $\mathbf{v} = (v^i) \in \mathbb{R}^3$, with $\|v\| < 1$. (Physically, $\mathbf{v}$ is the velocity of the second reference frame with respect to the first one.) The corresponding metric tensors $\breve{h}_1$ and $\breve{h}_2$, induced on B by the Bondi coordinate systems, are proportional to each other,

$$\breve{h}_2 = K^2 \breve{h}_1 , \tag{6.44}$$

where

$$K(z) := \frac{\sqrt{1 - \|v\|^2}}{1 - v_k z^k} = \frac{1 + v_k \bar{z}^k}{\sqrt{1 - \|v\|^2}} . \tag{6.45}$$

Consequently, the two-dimensional volume density induced on B by $\breve{h}$, represented in local coordinates x^A on S^2 by the odd form $\lambda = \sqrt{\det \breve{h}}\, dx^1 \wedge dx^2$, transforms according to the rule

$$\lambda \to \bar{\lambda} = K^2 \lambda . \tag{6.46}$$

On the other hand, those BMS transformations which asymptote to this boost transformation when restricted to B, change the scale in each fiber of $\mathscr{I}^+$ according to the formula (see [25, 109])

$$u \to \bar{u} = K(u - \alpha) . \tag{6.47}$$

In the Appendices C.5 and C.6 we give the transformation rules for objects of interest under the transformations (6.47).

We wish to find a geometric object which encodes the above synchronization between the transformation law (6.46) for the two-dimensional volume density in B and the law (6.47) for the change of scale in the fibers. For this purpose consider the set of metric tensors in each fiber $\pi^{-1}(b)$ invariant with respect to affine translations. In Bondi coordinates on $\mathscr{I}^+$ every such tensor may be written as $\Gamma = \gamma(du)^2$, where γ is a real number. The collection of all these tensors is a vector bundle over B, which we denote by $\mathbb{J}$. Its fibers are one-dimensional and may be parameterized by the variable γ. The transformation law (6.47) implies the following transformation law for elements of $\mathbb{J}$:

$$\gamma(du)^2 = \gamma\ K^{-2}(d\bar{u})^2 = \bar{\gamma}(d\bar{u})^2 ,$$

or, simply

$$\bar{\gamma} = \gamma\ K^{-2} .$$

Consequently, the three-dimensional volume density on $\mathbb{J}$,

$$\rho := \widehat{\lambda} \wedge d\gamma = \widehat{\bar{\lambda}} \wedge d\bar{\gamma} = \bar{\rho} ,$$

remains invariant under boost transformations. Here $\widehat{\lambda}$, respectively $\widehat{\bar{\lambda}}$, denotes the lift of λ, respectively of $\bar{\lambda}$, to $\mathbb{J}$. If we consider the set of metrics γdu^2 as a subset of the set of two-covariant tensors in each fiber $\pi^{-1}(b)$ invariant with respect to affine translations, then the vector structure of this last set implies the existence of a natural translation operation there. We note that the volume form ρ above is invariant under those translations; we shall say that it is *compatible with the vector structure of the fibers*. This is of course a slight abuse of terminology, since there is no vector structure on the fibers of $\mathbb{J}$ as such.

We are ready now to define the *abstract Scri space* $\mathscr{I}^+$, as being the structure $(\mathscr{I}^+, \pi, B, \mathscr{C}, \rho)$, where $(\mathscr{I}^+, \pi, B)$ is an affine bundle over B with one-dimensional oriented fibers, $\mathscr{C}$ is a conformal structure in B, and ρ is a volume density in $\mathbb{J}$, compatible with the vector structure of the fibers. We also assume that B is diffeomorphic to S^2.

So far we have shown how to obtain the abstract Scri structure starting from Bondi coordinates; let us show now how to obtain Bondi coordinate systems on $\mathscr{I}^+$, using the abstract structure $(\mathscr{I}^+, \pi, B, \mathscr{C}, \rho)$. In order to do this, consider $(S^2, \mathbf{h})$ equipped with the standard (unit, round) metric $\breve{h}$; a *standard parameterization* of $(B, \mathscr{C})$ by $(S^2, \breve{h})$ is a diffeomorphism

$$R : B \to S^2 , \tag{6.48}$$

such that the pull-back metric $R^*\breve{h}$ on B is compatible with the conformal structure of B. (It follows from the uniformization theorem that the set of standard parameterizations is not empty.) Each such parameterization will be called a *reference frame on* B. The volume form ρ on $\mathbb{J}$, compatible with the vector structure of the fibers, is precisely what we need to reconstruct uniquely the scale factor in the fibers of $\mathscr{I}^+$, once a reference frame R in B is chosen. This proceeds as follows: Consider the (two-dimensional) measure λ on B defined as above by the reference frame R. In each fiber $\mathbb{J}_p$ of $\mathbb{J}$ there is a unique (one-dimensional) measure μ_p which reproduces the three-dimensional measure ρ when multiplied by λ; one can think of μ_p as of $\frac{\rho}{\lambda}$, and this is actually defined as follows: Let $\widehat{\lambda}$ be the lift of λ to $\mathbb{J}$, and for $p \in B$ let $j_p : \mathbb{J}_p \to \mathbb{J}$ be the inclusion map of the fibers $\mathbb{J}_p$, projecting on p, of $\mathbb{J}$. Let $\hat{\mu}$ be any odd one-form field on $\mathbb{J}$ such that

$$\rho = \hat{\mu} \wedge \hat{\lambda} \,, \tag{6.49}$$

set

$$\mu_p = j_p^* \hat{\mu} \,.$$

One sees that μ_p is independent of the choice of $\hat{\mu}$ satisfying (6.49), and that μ_p is compatible with the vector structure because ρ is. Then, in each fiber of $\mathbb{J}$, we define the (positive) metric Γ_0 to be that which is at unit distance — according to the measure μ — from the null tensor $\{\gamma = 0\}$. Finally, among all possible affine parameters u in the fibers of $\mathscr{I}^+$ we choose the one which is compatible with the orientation and trivializes Γ_0, in the sense that

$$\Gamma_0 = (du)^2 \,. \tag{6.50}$$

It should be clear from what has been said above that the symmetry group of any abstract Scri structure $(\mathscr{I}^+, \pi, B, \mathscr{C}, \rho)$ is the BMS group.

6.4 Lorentz charges

Consider a hyperboloidal hypersurface $\mathscr{S}$, and choose a Bondi coordinate system so that $\partial\mathscr{S}$ is given by the equation

$$\partial\mathscr{S} = \{u = 0\} \subset \mathscr{I} \,. \tag{6.51}$$

This can always be achieved by performing a translation or supertranslation. Let X be a vector field satisfying the requirements spelled out at the beginning of Sect. 6.1; the integrand of the boundary term in the variational formula (5.27) takes then the form

$$16\pi \left(X^0 p^3{}_{\mu\nu} \delta \mathfrak{g}^{\mu\nu} - X^3 p^0{}_{\mu\nu} \delta \mathfrak{g}^{\mu\nu} \right) = -\frac{1}{2} \sin\theta \left[x^{-1} X^x \chi_{AB} + X^0 \partial_u \chi_{AB} \right] \delta\chi^{AB} \tag{6.52}$$

where, as before, $x^{-1}X^x$ at $x = 0$ is understood by a limiting process. Now, the boundary $\partial\mathscr{S}$ of every hyperboloidal hypersurface $\mathscr{S}$ singles out a preferred set of BMS generators by the requirement that X be tangent to $\partial\mathscr{S} \subset \mathscr{I}$. Recall that every generator of the BMS group is of the form

$$X = X_{boost} + X_{rot} + X_{supertr} \ ,$$

cf. Equations (6.10), (6.11) and (6.28); here translations are understood as a special case of supertranslations. It follows from (6.51) and (6.28) that X's tangent to $\partial\mathscr{S}$ have no supertranslation component, so that the variational formula (5.21) reads

$$-\delta H(X,\mathscr{S}) = \int_{\mathscr{S}} (\pounds_X p^\lambda{}_{\mu\nu}\delta\mathfrak{g}^{\mu\nu} - \pounds_X \mathfrak{g}^{\mu\nu}\delta p^\lambda{}_{\mu\nu})\, dS_\lambda$$
$$+\frac{1}{32\pi}\int_{S^2} \sin\theta x^{-1} X^x \chi_{AB}\delta\chi^{AB}\, d\theta\, d\varphi \ . \qquad (6.53)$$

This formula contains the required "$\dot{p}dq - \dot{q}dp$" term, together with the undesirable, non-Hamiltonian, boundary term which does not vanish in general unless $\lim_{x\to 0} x^{-1}X^x$ does. This problem is easily cured by replacing H with

$$\hat{H}_L(X,\mathscr{S}) = H(X,\mathscr{S}) + \frac{1}{64\pi}\int_{S^2} \sin\theta x^{-1} X^x \chi_{AB}\chi^{AB}\, d\theta\, d\varphi \ , \qquad (6.54)$$

so that we obtain

$$-\delta \hat{H}_L(X,\mathscr{S}) = \int_{\mathscr{S}} (\pounds_X p^\lambda{}_{\mu\nu}\delta\mathfrak{g}^{\mu\nu} - \pounds_X \mathfrak{g}^{\mu\nu}\delta p^\lambda{}_{\mu\nu})\, dS_\lambda \ . \qquad (6.55)$$

We emphasize that the convergence of all the objects at hand has already been established in the previous sections. equation (6.55) shows that if one considers only those vector fields X which are tangent to $\partial\mathscr{S}$, then the complications related to the inclusion in the phase spaces of the "c data" on $\mathscr{I}$ can be avoided. In fact, it is sufficient to consider the usual phase space (γ,π) of fields on $\mathscr{S}$ such that the associated conformal data $((\tilde{\gamma},\omega),(\tilde{\pi},\dot{\omega}))$ satisfy the requirements spelled out in Sect. 5.7. It follows from equation (6.55) that the $\hat{H}_L(X,\mathscr{S})$'s are Hamiltonians on this space.

For $\tau \geq -1$ let the hypersurfaces $\mathscr{S}_\tau$ be as in Sect. 5.7; equation (6.55) further shows that the $\hat{H}_L$'s are Hamiltonians on the phase space $\mathscr{P}_{[-1,0]}$ of that section, and on the phase space $\widehat{\mathscr{P}}_{[-1,0]}$ of Sect. 5.8, for rotations, as well as for those boosts which are *tangent to* $\partial\mathscr{S}_{-1}$: indeed, from what has been said one easily finds that for all $\tau_* \geq -1$ we have

$$-\delta \hat{H}_L(X,\mathscr{S}_{-1}) = \int_{\mathscr{S}_{\tau_*}} (\pounds_X p^\lambda{}_{\mu\nu}\delta\mathfrak{g}^{\mu\nu} - \pounds_X \mathfrak{g}^{\mu\nu}\delta p^\lambda{}_{\mu\nu})\, dS_\lambda$$
$$+\int_{-1}^{\tau_*} d\tau \int_{S_\tau} \left(\frac{\partial\pounds_X\chi_{AB}}{\partial u}\delta\chi^{AB} - \pounds_X\chi^{AB}\frac{\partial\delta\chi_{AB}}{\partial u}\right)\frac{\sin\theta}{32\pi}\, d\theta\, d\varphi \ ;$$
$$(6.56)$$

compare equation (5.128).

So far the vector fields X in (6.54) were Killing vector fields of the background in the asymptotic region; if they are so globally, then the Hamiltonians $\hat{H}_L(X,\mathscr{S})$ coincide with the $H_L(X,\mathscr{S})$'s defined as

$$H_L(X,\mathscr{S}) = H_{\text{boundary}}(X,\mathscr{S}) + \frac{1}{64\pi}\int_{S^2} \sin\theta x^{-1} X^x \chi_{AB}\chi^{AB}\, d\theta\, d\varphi\,. \quad (6.57)$$

The H_L's provide Hamiltonians for ADM-type variational formulae: From (5.43) and (5.121) one similarly obtains, for X's tangent to $\partial\mathscr{S}$:

$$-\delta H_L(X,\mathscr{S}) = \frac{1}{16\pi}\int_\Sigma \left(\dot{\pi}^{kl}\delta\gamma_{kl} - \dot{\gamma}_{kl}\delta\pi^{kl}\right) d^3x\,. \quad (6.58)$$

Equation (6.58) and the associated Hamiltonians (6.57) seem to have a more fundamental character than (6.55) and (6.54), because no background is explicitly needed in the right-hand-side of (6.58); we shall therefore use H_L rather than $\hat{H}_L$ in our proposal below how to define the global Lorentz charges of cuts of $\mathscr{I}^+$.

6.5 A Hamiltonian definition of angular momentum of sections of $\mathscr{I}$

In this section we wish to present a method to handle the "angular momentum ambiguities". In our context this problem can be formulated as follows: in order to calculate the Hamiltonians in the formalism presented above one needs to prescribe: 1) some background metric b, and 2) a vector field X. The asymptotic conditions imposed single out a preferred family of X's: the BMS Lie algebra. By definition, this is the Lie algebra of boosts, rotations, translations and supertranslations. Setting aside the question of choice of the background metric, there is no known geometric way to distinguish between two BMS vector fields which both asymptote to the same rotation, or the same boost; in other words, there is no known geometric way of singling out the boost and rotation generators in the BMS algebra. Therefore there is no unique way of deciding, which vector field X should be used in the integrals (6.54) or (6.57) when attempting to define the Lorentz group charges — that is, the angular momentum and boost charges.

A way out of this problem is suggested by the results of Sect. 6.4, using the fact that in a Hamiltonian framework one assigns global Lorentz group charges *to a hypersurface* $\mathscr{S}$. Let then a hypersurface $\mathscr{S}$ be given, and suppose that $\mathscr{S}$ intersects $\mathscr{I}^+$ transversally at a smooth section $S = \partial\mathscr{S} = \mathscr{S}\cap\mathscr{I}^+$. Such a section singles out a six parameter family of Bondi coordinate systems, by the requirement that in the chosen Bondi coordinates we have $S = \{u = 0\}$. Now, every such coordinate system defines a flat background metric b in a neighbourhood of S via equation (5.96). The resulting metrics are independent of the Bondi coordinate system chosen, within the six parameter freedom available, as those coordinate systems differ from each other by a Lorentz transformation. We can thus define a unique six parameter family of BMS generators which are singled out by the requirement that they are tangent to S, and that they are Killing vector fields of the background metric b. This is similar to the approach in Sect. 6.4; however, here we emphasize the fact that $\partial\mathscr{S}$ defines a naturally preferred background b.

The above singling out of the vector fields X appears to be quite natural from a symplectic point of view, because — as shown in detail in Sect. 6.4 — for vector fields which are tangent to $\partial\mathscr{S}$ the boundary integral in the variational formula (6.55) (which is equivalent to the more usual Arnowitt-Deser-Misner (ADM) type formula (6.58)) vanishes. This isn't true for the remaining BMS generators in general. One is thus tempted to use equation (6.57) with the above constructed background metric and vector fields to define the relevant global Lorentz charges. From (6.14) one obtains with some work [83, pp. 715 and 728]

$$16\pi \int_{S_\tau} \mathbb{W}^{xu}\Big|_{\mathscr{I}^+} d\theta\, d\varphi = -6 \int_{S_\tau} \breve{N}_A X^A \sin\theta\, d\theta\, d\varphi\,, \tag{6.59}$$

where

$$\breve{N}_A := N_A + \frac{1}{12}\chi_{AB}\chi^{BC}{}_{||C}\,, \tag{6.60}$$

hence

$$H_L(X,\mathscr{S}) = -\frac{1}{64\pi}\int_{S^2}\left(24\breve{N}_A X^A\big|_{x=0} + \chi_{AB}\chi^{AB}\frac{\partial X^x}{\partial x}\big|_{x=0}\right)\sin\theta\, d\theta\, d\varphi\,. \tag{6.61}$$

Formula (6.61) is our proposal how to assign global Lorentz charges to hyperboloidal hypersurfaces. We emphasize that the background b used above to calculate N_A, as well as χ_{AB}, is uniquely defined by the metric g and the section $\partial\mathscr{S}$ of $\mathscr{I}$, and that the vector fields X in (6.61) should belong to the six dimensional vector space of BMS generators uniquely singled out by $\partial\mathscr{S}$.

It is of interest to apply our proposal to the Kerr metric. Inserting Equation (C.130) of Appendix C.7 in the explicit formulae (6.120)-(6.121) for H_L, derived below, gives the expected answer for the Lorentz charges of sections $u = \text{const}$ of $\mathscr{I}^+$, where u is a Bondi coordinate system asymptotic to the Eddington-Finkelstein one. The reader is referred to Sect. 6.7 below for the calculation of the Lorentz charges on other sections of $\mathscr{I}^+$ for general stationary space-times, which applies in particular to the Kerr metric.

One would expect the correct definition of Lorentz group charges to produce zero value on any section of $\mathscr{I}^+$ for Minkowski space-time. Now, in Minkowski space-time, for general sections of $\mathscr{I}^+$ the background metric so obtained will be related to the standard one by a non-trivial supertranslation. Let the section S of $\mathscr{I}^+$ be given by the equation $\{u = \lambda\}$ in a standard Minkowskian coordinate system, and let $\breve{N}^A$ be the function (6.60) for the Minkowski metric in a Bondi coordinate system, which differs from the standard Minkowskian one by the coordinate transformation (C.107)–(C.109); from the definition (6.60), the transformation formula (C.118) of Appendix C.5, and the identity

$$\lambda_{||AB}{}^{B} = (\Delta_2\lambda)_{||A} + \lambda_{||A} \,, \tag{6.62}$$

one finds

$$\breve{N}^A = \frac{1}{12}\left(2\lambda^{||A}{}_B - \delta^A_B\Delta_2\lambda\right)[(\Delta_2+2)\lambda]^{||B} \,. \tag{6.63}$$

Let, first, X be a rotational Killing vector field, equation (6.10) at $x = 0$ gives

$$X^A = \epsilon^{AB}\partial_B v \,, \qquad \frac{\partial X^x}{\partial x} = 0 \,, \tag{6.64}$$

thus only the $\breve{N}_A$ term matters in this case. Inserting (6.63)-(6.64) into (6.59), after some straightforward rearrangements and integrations by parts one has

$$12\int_{S^2}\breve{N}_A\varepsilon^{AB}v_{||B} = \int_{S^2}\left(2\lambda_{||AB} - \breve{h}_{AB}\Delta_2\lambda\right)[(\Delta_2+2)\lambda]^{||B}\,\varepsilon^{AC}v_{||C}$$

$$= -2\int_{S^2}\lambda_{||AB}{}^{B}\,[(\Delta_2+2)\lambda]\,\varepsilon^{AC}v_{||C} \tag{6.65}$$

$$-2\int_{S^2}\lambda_{||AB}\,[(\Delta_2+2)\lambda]\,\varepsilon^{AC}v_{||C}{}^{B} \tag{6.66}$$

$$-\int_{S^2}(\Delta_2+2)\lambda\,[(\Delta_2+2)\lambda]_{||A}\,\varepsilon^{AC}v_{||C} \tag{6.67}$$

$$-2\int_{S^2}\lambda_{||A}\,[(\Delta_2+2)\lambda]\,\varepsilon^{AC}v_{||C} \,. \tag{6.68}$$

Now, v is a linear combination of the $\ell = 1$ spherical harmonics and therefore satisfies the identity

$$v_{||AB} = -\breve{h}_{AB}v \,, \tag{6.69}$$

hence

$$\lambda_{||AB}\varepsilon^{AC}v_{||C}{}^{B} = -v\lambda_{||AB}\varepsilon^{AB} = 0 \,,$$

so that the term (6.66) vanishes. The term (6.67) gives no contribution either because

$$\int_{S^2}(\Delta_2+2)\lambda\,[(\Delta_2+2)\lambda]_{||A}\,\varepsilon^{AC}v_{||C} = \frac{1}{2}\int_{S^2}\left[((\Delta_2+2)\lambda)^2\right]_{||A}\varepsilon^{AC}v_{||C}$$

$$= -\frac{1}{2}\int_{S^2}\left[((\Delta_2+2)\lambda)^2\right]\varepsilon^{AC}v_{||CA} = 0 \,, \tag{6.70}$$

and so we are left with the sum of the first and last terms (6.65) and (6.68):

$$2\int_{S^2}\left(\lambda_{||AB}{}^{B} + \lambda_{||A}\right)[(\Delta_2+2)\lambda]\,\varepsilon^{AC}v_{||C} = \int_{S^2}\left[((\Delta_2+2)\lambda)^2\right]_{||A}\varepsilon^{AC}v_{||C}$$
$$= 0 \,.$$

Here we have used (6.62) together with (6.70). Now, Equation (C.124) shows that $N^A = 0$, hence $N^A = \breve{N}^A$, and finally

$$H_L(\epsilon^{AB}\partial_B v, \mathscr{S}) = -\frac{3}{8\pi}\int_{S^2} \breve{N}_A \varepsilon^{AB} v_{||B} \sin\theta \, d\theta \, d\varphi = 0 \,, \tag{6.71}$$

as desired. Consider, next, boost vector fields; equation (6.11) at $x = 0$ gives

$$X^A = \breve{h}^{AB}\partial_B v \,, \qquad \frac{\partial X^x}{\partial x} = -v \,. \tag{6.72}$$

The first integral in (6.61) can then be manipulated as follows:

$$\begin{aligned} 12\int_{S^2} \breve{N}_A \breve{h}^{AB} v_{||B} &= \int_{S^2} \left(2\lambda_{||AB} - \breve{h}_{AB}\Delta_2\lambda\right) [(\Delta_2+2)\lambda]^{||B} v^{||A} \\ &= -2\int_{S^2} \lambda_{||AB}{}^{B} \, [(\Delta_2+2)\lambda] \, v^{||A} \\ &\quad -2\int_{S^2} \lambda_{||AB} \, [(\Delta_2+2)\lambda] \, v^{||AB} \\ &\quad -\int_{S^2} \Delta_2\lambda \, [(\Delta_2+2)\lambda]_{||A} \, v^{||A} \,; \end{aligned} \tag{6.73}$$

an integration by parts has been performed. Equation (6.69) leads to

$$\lambda_{||AB} v^{||AB} = -v\Delta_2\lambda \,.$$

This, the commutation identity

$$\lambda_{||AB}{}^{B} = (\Delta_2\lambda)_{||A} + \lambda_{||A} \,,$$

some rearrangements and integrations by parts allow us to rewrite (6.73) as follows:

$$\begin{aligned} &\int_{S^2} \left(2\lambda_{||AB} - \breve{h}_{AB}\Delta_2\lambda\right) [(\Delta_2+2)\lambda]^{||B} v^{||A} \\ &= -\int_{S^2} [(\Delta_2+2)\lambda]_{||A} \, [(\Delta_2+2)\lambda] \, v^{||A} \\ &\quad + 2\int_{S^2} \Delta_2\lambda \, [(\Delta_2+2)\lambda] \, v - \int_{S^2} [\Delta_2\lambda(\Delta_2+2)\lambda]_{||A} \, v^{||A} \\ &= -\frac{1}{2}\int_{S^2} \left[((\Delta_2+2)\lambda)^2\right]_{||A} v^{||A} + \int_{S^2} \Delta_2\lambda \, [(\Delta_2+2)\lambda] \, (\Delta_2+2)v \\ &= \frac{1}{2}\int_{S^2} \left[((\Delta_2+2)\lambda)^2\right] \Delta_2 v \,, \end{aligned} \tag{6.74}$$

in the last step we have used the fact that $(\Delta_2+2)v = 0$. We thus have

$$\int_{S^2} \breve{N}^A v_{||A} \sin\theta \, d\theta \, d\varphi = -\frac{1}{12}\int_{S^2} [(\Delta_2+2)\lambda]^2 \, v \, \sin\theta \, d\theta \, d\varphi \,. \tag{6.75}$$

Very similar manipulations show that the $\partial X^x/\partial x$ term cancels precisely the contribution (6.75) from the X^A terms leading again to a vanishing value of $H_L(X,\mathscr{S})$, for all hyperboloidal hypersurfaces in Minkowski space-time, and for all X's chosen as prescribed above; the details proceed as follows: First, a few integrations by parts lead to the identity

$$\begin{aligned}
\int_{S^2} \lambda_{||A} v^{||A} \Delta_2 \lambda &= -\int_{S^2} (\lambda_{||A} v^{||A})_{||B} \lambda^{||B} \\
&= -\int_{S^2} \lambda_{||BA} v^{||A} \lambda^{||B} - \int_{S^2} v^{||AB} \lambda_{||A} \lambda_{||B} \\
&= -\frac{1}{2}\int_{S^2} (\lambda_{||B}\lambda^{||B})_{||A} v^{||A} + \int_{S^2} v \lambda^{||A} \lambda_{||A} \\
&= \frac{1}{2}\int_{S^2} (\lambda_{||B}\lambda^{||B})(\Delta_2 + 2) v = 0 \,, \qquad (6.76)
\end{aligned}$$

Next,

$$\begin{aligned}
\int_{S^2} v \lambda_{||AB} \lambda^{||AB} &= -\int_{S^2} (v \lambda_{||AB})^{||B} \lambda^{||A} \\
&= -\int_{S^2} v^{||B} \lambda_{||BA} \lambda^{||A} - \int_{S^2} v \lambda_{||AB}{}^{||B} \lambda^{||A} \qquad (6.77) \\
&= \int_{S^2} (v^{||B} \lambda^{||A})_{||A} \lambda_{||B} - \int_{S^2} v \left[(\Delta_2 + 1)\lambda\right]_{||A} \lambda^{||A} \qquad (6.78) \\
&= \int_{S^2} v^{||BA} \lambda_{||B} \lambda_{||A} + \int_{S^2} v^{||B} \Delta_2 \lambda \lambda_{||B} - \int_{S^2} v \lambda_{||A} \lambda^{||A} \\
&\quad + \int_{S^2} v_{||A} \left[\Delta_2 \lambda\right] \lambda^{||A} + \int_{S^2} v \left[\Delta_2 \lambda\right] \Delta_2 \lambda \qquad (6.79) \\
&= \int_{S^2} v \left[(\Delta_2 \lambda)^2 - 2 \lambda^{||A} \lambda_{||A}\right] . \qquad (6.80)
\end{aligned}$$

In passing from (6.77) to (6.78) an integration by parts on the first term has been performed; the third term in (6.79) arises from the last one in (6.78); the second and fourth term in (6.79) vanish by (6.76). Further,

$$\begin{aligned}
\int_{S^2} v \lambda^{||A} \lambda_{||A} &= -\int_{S^2} v \lambda \Delta_2 \lambda - \int_{S^2} v_{||A} \lambda^{||A} \lambda \\
&= -\int_{S^2} v \lambda \Delta_2 \lambda - \frac{1}{2}\int_{S^2} v_{||A} \left(\lambda^2\right)^{||A} \\
&= -\int_{S^2} v \lambda \Delta_2 \lambda + \frac{1}{2}\int_{S^2} (\Delta_2 v) \lambda^2 \\
&= -\int_{S^2} v \lambda (\Delta_2 + 1) \lambda \,; \qquad (6.81)
\end{aligned}$$

in the last step, Equation (6.69) has been used. Equations (6.80)-(6.81) lead to

$$\begin{aligned}\int_{S^2} v\chi_{AB}\chi^{AB} &= \int_{S^2}\left(2\lambda_{||AB}-\breve{h}_{AB}\Delta_2\lambda\right)\left(2\lambda^{||AB}-\breve{h}^{AB}\Delta_2\lambda\right)v \quad (6.82)\\ &= 4\int_{S^2} v\lambda_{||AB}\lambda^{||AB} - 2\int_{S^2} v(\Delta_2\lambda)^2\\ &= 2\int_{S^2} v\left[(\Delta_2\lambda)^2 - 4\lambda^{||A}\lambda_{||A}\right]\\ &= 2\int_{S^2} v\left[(\Delta_2\lambda)^2 + 4\lambda(\Delta_2+1)\lambda\right]\\ &= 2\int_{S^2} v\left[(\Delta_2+2)\lambda\right]^2 ,\end{aligned}$$

and finally

$$\begin{aligned}-\frac{1}{4}\int_{S^2}\chi_{AB}\chi^{AB}\frac{\partial X^x}{\partial x}\Big|_{x=0}\sin\theta\, d\theta\, d\varphi &= \frac{1}{4}\int_{S^2} v\chi_{AB}\chi^{AB}\sin\theta\, d\theta\, d\varphi\\ &= \frac{1}{2}\int_{S^2} v\left[(\Delta_2+2)\lambda\right]^2\sin\theta\, d\theta\, d\varphi \quad (6.83)\end{aligned}$$

equations (6.61), (6.75) and (6.83) prove the desired vanishing of the boost charges.

In the next section we show that the Lorentz charges so defined have all the properties one might expect for sections of $\mathscr{I}^+$ in Schwarzschild space-time.

For alternative approaches to the definition of angular momentum at null infinity and their problems, see [13, 29, 55, 117, 126] and references therein.

6.6 An example: Schwarzschild space-time

It is of interest to apply our definition of Lorentz charges to sections of Scri in the Schwarzschild space-time. The outgoing Eddington-Finkelstein coordinates [100, p. 830] provide us with a natural Bondi coordinate system

$$\begin{aligned}g_{\mu\nu}dx^\mu dx^\nu &\equiv -(1-\frac{2m}{r})du^2 - 2du\, dr + r^2\breve{h}_{AB}dx^A dx^B\\ &= -(1-2mx)du^2 + 2x^{-2}du\, dx + x^{-2}\breve{h}_{AB}dx^A dx^B , \quad (6.84)\end{aligned}$$

so that the associated Bondi functions take the simple form

$$M = m\,, \quad \chi_{AB} = 0\,, \quad N^A = 0\,. \quad (6.85)$$

Let a section S of $\mathscr{I}^+ := \{x = 0\}$ be given by the equation

$$u = \alpha(\theta,\varphi)\,,$$

for some smooth function α. Let the new Bondi coordinate $\bar{u}$ be obtained from u by a translation or a supertranslation,

$$\bar{u}|_{x=0} = u - \alpha \; ; \tag{6.86}$$

using the formulae of Appendix C.5 one finds

$$\bar{M} = m \, , \quad \bar{\chi}_{AB} = -2\alpha_{||AB} + \breve{h}_{AB}\Delta_2\alpha \, , \quad \bar{N}_A = -m\alpha_{,A} \, . \tag{6.87}$$

equation (6.18) gives the expected formula

$$p^0 = m \, , \qquad p^i = 0 \, . \tag{6.88}$$

Recall that *rest frames* are usually defined by the condition $p^i = 0$; Equation (6.88) shows that the Bondi coordinates determined by the Eddington-Finkelstein outgoing coordinates, as well as those related to them by (6.86), correspond to an asymptotic rest frame for the Schwarzschild space-time, as expected.

Now, $\bar{\chi}_{AB}$ in (6.87) has exactly the same form as the corresponding object in Minkowski space-time, so that the calculation just done in the Minkowski case gives the identity

$$\int_{S^2} \left(2\bar{\chi}_{AB}\bar{\chi}^{BC}{}_{||C}X^A\big|_{x=0} + \bar{\chi}_{AB}\bar{\chi}^{AB}\frac{\partial X^x}{\partial x}\big|_{x=0} \right) \sin\theta \, d\theta \, d\varphi = 0 \tag{6.89}$$

equations (6.60) and (6.61) then give

$$\begin{aligned} H_L(X, \mathscr{S}) &= \frac{3m}{8\pi} \int_{S^2} X^A \alpha_{||A} \sin\theta \, d\theta \, d\varphi \\ &= -\frac{3m}{8\pi} \int_{S^2} X^A{}_{||A} \alpha \sin\theta \, d\theta \, d\varphi \, . \end{aligned} \tag{6.90}$$

Here $\mathscr{S}$ is any hypersurface in space-time such that

$$\partial\mathscr{S} \cap \mathscr{I}^+ = S \, .$$

For rotational Killing vectors (6.72) we have $X^A{}_{||A} = 0$, and Equation (6.90) shows that *our proposed Hamiltonian angular momentum vanishes for all cuts of $\mathscr{I}^+$ of the Schwarzschild space-time*, in the reference frame determined by the Eddington-Finkelstein outgoing coordinates.

Consider, next boost generators; we have $X^A{}_{boost} = v_{boost}^{||A}$, where v_{boost} is a linear combination of $\ell = 1$ spherical harmonics, and from (6.90) we obtain

$$\begin{aligned} H_L(X_{boost}, \mathscr{S}) &= -\frac{3m}{8\pi} \int_{S^2} (\Delta_2 v_{boost})\alpha \sin\theta \, d\theta \, d\varphi \\ &= \frac{3m}{4\pi} \int_{S^2} v_{boost}\alpha \sin\theta \, d\theta \, d\varphi \, . \end{aligned} \tag{6.91}$$

where we have used $(\Delta_2 + 2)v_{boost} = 0$. It follows that all the Hamiltonian Lorentz charges (6.57) of a cut S vanish if and only if v has vanishing $\ell = 1$

coefficients in a spherical harmonics expansion. This result has the following natural interpretation: Consider a relativistic field theory in Minkowski space-time with canonical energy-momentum tensor $T_{\mu\nu}$, so that

$$p^\mu = -\int_{t=0} T^{\mu 0} d^3x \ ,$$

where the minus sign originates[3] from our signature $(-,+,+,+)$. Let $L^{\mu\nu}$ be the usual global Lorentz charges given by

$$L^{\mu\nu} = -\int_{t=0} \left(x^\mu T^{\nu 0} - x^\nu T^{\mu 0} \right) d^3x \ ;$$

equivalently

$$\frac{1}{2} L^{\mu\nu} J_{\mu\nu} = -\int_{t=0} T^{\mu 0} X_\mu d^3x \ , \tag{6.92}$$

with

$$X \equiv X(J) := J_\mu{}^\nu x^\mu \partial_\nu \ , \tag{6.93}$$

where $J_{\mu\nu}$ is an arbitrary anti-symmetric matrix with constant coefficients. It follows that under a Minkowskian translation

$$x^\mu \longrightarrow \bar{x}^\mu = x^\mu + a^\mu \ , \tag{6.94}$$

where the a^μ's are constants, the $L^{\mu\nu}$'s transform as

$$L^{\mu\nu} \longrightarrow \bar{L}^{\mu\nu} = L^{\mu\nu} + a^\mu p^\nu - a^\nu p^\mu \ . \tag{6.95}$$

Now, a Minkowskian translation corresponds to a BMS generator with α given by

$$\alpha = a^0 - a_i n^i \ ,$$

cf. equations (6.9) and (6.16), which is a linear combination of the $\ell = 0$ and $\ell = 1$ spherical harmonics. For X's of the form (6.93) from (4.68) we obtain

$$v_{boost} = J^0{}_k n^k$$

so that

$$\begin{aligned} H_L(X(J), \mathscr{S}) &= \frac{3m}{4\pi} \int_{S^2} v_{boost} \alpha \sin\theta \, d\theta \, d\varphi \\ &= -m J^0{}_k a^k = a^{[\mu} p^{\nu]} J_{\mu\nu} \ , \end{aligned} \tag{6.96}$$

and we have used

[3] In our conventions the canonical energy momentum tensor for a massless scalar field satisfies $T^0{}_0 \geq 0$, and equals *minus* the energy momentum tensor appearing as a source term in the Einstein equations minimally coupled to such a field.

$$\frac{1}{4\pi}\int_{S^2} n^k n_l \sin\theta \, d\theta \, d\varphi = \frac{1}{3}\delta^k{}_l \, .$$

This is identical with the transformation law (6.95).

Any α in (6.86) can be decomposed as

$$\alpha = \alpha_{\rm trans} + \alpha_{\rm supertrans} \, ,$$

where $\alpha_{\rm supertrans}$ has no $\ell = 0$ or $\ell = 1$ spherical harmonics; here we are momentarily suspending our convention that translations are a special case of supertranslations. Similarly the coordinate change (6.86) can be viewed as a composition of a first coordinate change in which α is replaced by $\alpha_{\rm trans}$, followed by a second one with α replaced by $\alpha_{\rm supertrans}$. The first transformation leads to sections of $\mathscr{I}^+$ with Lorentz charges given by (6.96); somewhat surprisingly, the second one does not change the value of those charges.

The global Lorentz charges with respect to moving frames — *i.e.*, Bondi coordinates boosted with respect to the ones above — can be obtained using the transformation properties derived in Section 6.8 below.

6.7 An example: stationary space-times

The next example to which we apply our definition of global Lorentz charges is that of stationary space-times. Building upon the work of Beig and Simon [22], Damour and Schmidt [52] have derived[4] the asymptotic form of the metric near $\mathscr{I}^+$ for stationary vacuum space-times: in the coordinate system of [52, Appendix] those metrics take the following form

$$\begin{aligned} g = & -(A^2 + O(\tilde r^{-3}))d\tilde u^2 - 2(AB + O(\tilde r^{-3}))d\tilde u d\tilde r + 2\psi d\tilde\varphi(d\tilde u + \frac{B}{A}d\tilde r) \\ & + O(\tilde r^{-2})d\tilde u d\tilde v^A + O(\tilde r^{-3})d\tilde r^2 + O(\tilde r^{-2})d\tilde r d\tilde v^A \\ & + B^2\tilde r^2\left(\breve h_{AB} + O(r^{-3})\right)d\tilde v^A d\tilde v^B \, , \end{aligned} \tag{6.97}$$

with $\tilde v^A = (\tilde\theta, \tilde\varphi)$ — standard angular coordinates on a sphere. Here

$$A^2 = 1 - \frac{2m}{r} + \frac{2m^2}{r^2} \, , \quad B^2 = 1 + \frac{2m}{r} + \frac{3m^2}{2r^2} \, , \quad \psi = -\frac{2ma\sin^2\tilde\theta}{\tilde r} \, , \tag{6.98}$$

where m and a are constants, It follows that

$$\begin{aligned} & g^{\tilde u\tilde u} = \tfrac{\psi^2}{r^2\sin^2\theta} + O(\tilde r^{-5}) \, , \quad g^{\tilde u\tilde r} = -\tfrac{1}{AB} + O(\tilde r^{-3}) \, , \\ & g^{\tilde u\tilde\varphi} = \tfrac{\psi}{r^2\sin^2\theta} + O(\tilde r^{-4}) \, , \\ & g^{\tilde u\tilde\theta} = O(\tilde r^{-4}) \, , \quad g^{\tilde r\tilde r} = B^{-2} + O(\tilde r^{-3}) \, , \quad g^{\tilde r\tilde\varphi} = O(\tilde r^{-4}) \, , \quad g^{\tilde r\tilde\theta} = O(\tilde r^{-4}) \, , \\ & g^{\tilde\varphi\tilde\varphi} = \tfrac{1}{B^2\tilde r^2\sin^2\tilde\theta} + O(\tilde r^{-5}) \, , \quad g^{\tilde\varphi\tilde\theta} = O(\tilde r^{-5}) \, , \quad g^{\tilde\theta\tilde\theta} = \tfrac{1}{B^2\tilde r^2} + O(\tilde r^{-5}) \, . \end{aligned} \tag{6.99}$$

[4] A gap in the work of [52] has been recently filled by Dain [50].

Using (6.97)-(6.99) it is simple to show that there exists a Bondi coordinate system (u, r, θ, φ) such that

$$u - \tilde{u} = O(\tilde{r}^{-3}) , \quad r - B\tilde{r} = O(\tilde{r}^{-2}) , \quad \varphi - \tilde{\varphi} = O(\tilde{r}^{-2}) , \quad \theta - \tilde{\theta} = O(\tilde{r}^{-3}) , \tag{6.100}$$

with the error terms behaving in the obvious way under differentiation. This leads immediately to

$$\chi_{AB} = 0 , \quad M = m . \tag{6.101}$$

To find N^A some more work is required. First, one finds the following asymptotic expansions

$$u - \tilde{u} = -\frac{2m^2a^2}{3\tilde{r}^3} + O(\tilde{r}^{-4}) , \tag{6.102}$$

$$\varphi - \tilde{\varphi} = \frac{ma}{\tilde{r}^2} + O(\tilde{r}^{-3}) , \tag{6.103}$$

The function N^A can now be obtained from the asymptotic expansion (5.99) for U^A together with the formula

$$U^A = -e^{2\beta} g^{rA}$$

(compare Equation (C.7)). A straightforward calculation of $g^{rA} := g(dr, dv^A)$, using Equations (6.99)-(6.103), proves that the leading order behaviour of the associated Bondi functions is identical to that for the Kerr metric, as in Equation (C.130), Appendix C.7:

$$\chi_{AB} = 0 = N^\theta , \quad M = m , \quad N^\varphi = ma . \tag{6.104}$$

This result remains true for stationary space-times which are not vacuum, provided the energy momentum tensor falls off sufficiently fast. We can repeat step by step the analysis already done for the Schwarzschild metric in Section 6.6: Equation (6.87) becomes

$$\bar{M} = m , \quad \bar{\chi}_{AB} = -2\alpha_{||AB} + \breve{h}_{AB}\Delta_2\alpha , \quad \bar{N}_A = N_A - m\alpha_{,A} , \tag{6.105}$$

so that (6.88) still holds:

$$p^0 = m , \qquad p^i = 0 . \tag{6.106}$$

Equation (6.89) remains valid, while (6.90) is replaced by

$$\begin{aligned} H_L(X, \mathscr{S}) &= -\frac{3}{8\pi} \int_{S^2} X^A (N_A - m\alpha_{||A}) \sin\theta \, d\theta \, d\varphi \\ &= -\frac{3}{8\pi} \int_{S^2} (X^A N_A + m X^A{}_{||A}\alpha) \sin\theta \, d\theta \, d\varphi . \end{aligned} \tag{6.107}$$

Here, as before, $\mathscr{S}$ is any hypersurface in space-time such that

$$\partial\mathscr{S} \cap \mathscr{I}^+ = S .$$

For rotational Killing vectors (6.72) we have $X^A{}_{||A} = 0$, and Equation (6.107) shows that *our proposed Hamiltonian angular momentum does not depend upon the cut of $\mathscr{I}^+$ for all stationary asymptotically vacuum space-times*, in the reference frame determined by the Damour-Schmidt coordinates. (Recall that the splitting of Lorentz generators into rotations and boosts requires the choice of a reference frame.) From what has been said it follows that it takes the expected value on all cuts of $\mathscr{I}$ in Kerr space-time.

Consider, next boost generators, for which $X^A{}_{boost} = v_{boost}^{||A}$, where v_{boost} is a linear combination of $\ell = 1$ spherical harmonics. From (6.107) we obtain

$$\begin{aligned} H_L(X_{boost}, \mathscr{S}) &= -\frac{3}{8\pi}\int_{S^2} (v_{boost}^{||A} N_A + m\Delta_2 v_{boost}\alpha) \sin\theta \, d\theta \, d\varphi \\ &= \frac{3}{8\pi}\int_{S^2} v_{boost}(N^A{}_{||A} + 2m\alpha) \sin\theta \, d\theta \, d\varphi \\ &= \frac{3m}{4\pi}\int_{S^2} v_{boost}\alpha \sin\theta \, d\theta \, d\varphi \, . \end{aligned} \tag{6.108}$$

In the first step above we have used $(\Delta_2 + 2)v_{boost} = 0$ in the last term, integrated by parts in the first term; in the second we have used the fact that $N^A\partial_A = ma\partial_\varphi$ given by (6.104) is a Killing vector on S^2, hence satisfies $N^A{}_{||A} = 0$. It follows that all the boost charges (6.57) — with the splitting into rotations and boosts determined by the Damour-Schmidt coordinate system — of a cut of $\mathscr{I}$ vanish if and only if v has vanishing $\ell = 1$ coefficients in a spherical harmonics expansion. Equations (6.107)-(6.108) show that the discussion of the behaviour of the Lorentz charges under supertranslations is identical to the one for Schwarzschild space-time, in particular we recover equation (6.95) for translations, and we recover the fact that pure supertranslations do not change the numerical values of the global Lorentz charges.

6.8 Lorentz covariance of global charges

In this section we wish to analyse the behaviour, under Lorentz transformations, of the global charges associated with the Poincaré group. Throughout this section we will suppose that X is a Killing vector of the background metric; what has been said so far shows that for such X's the Hamiltonian $H(X, \mathscr{S})$ depends only upon $\partial\mathscr{S}$. For our purposes here it is convenient to change notation, and to write

$$H(X, \partial\mathscr{S}, g, b)$$

for $H(X, \mathscr{S})$, emphasising thus the dependence of the Hamiltonian upon the metric g and the background metric b.

Let $\Psi : \mathcal{O} \to M$ be any smooth diffeomorphism defined on a neighbourhood $\mathcal{O}$ of ∂S, since our formula for the Hamiltonian $H(X, \mathscr{S})$ is diffeomorphism-invariant we have

$$H(\Psi_* X, \Psi(S), g, b) = H(X, S, \Psi^* g, \Psi^* b) , \tag{6.109}$$

where Ψ^* is the pull-back, and Ψ_* is the push-forward map. If one thinks of Ψ as a change of coordinates,

$$\bar{y}^\mu = \Psi^\mu(y^\nu) ,$$

and $\Psi_* X, g, b$ are given in the barred coordinate system $\bar{y}^\mu$, then $X, \Psi^* g, \Psi^* b$ can be thought of as the same objects $\Psi_* X, g, b$ expressed in the unbarred coordinates y^ν.

Consider, now, the flat background metric

$$b = -du^2 + 2dudr + r^2\left((d\theta)^2 + \sin^2\theta (d\varphi)^2\right) ;$$

b can be written in the standard Minkowskian form diag$(-1, +1, +1, +1)$ by setting

$$t = u + r ,$$
$$y^1 = r \sin\theta \cos\varphi , \quad y^2 = r \sin\theta \sin\varphi , \quad y^3 = r\cos\theta . \tag{6.110}$$

Let $\Psi(\Lambda)$ be a boost transformation,

$$\bar{y}^\mu = \Lambda^\mu{}_\nu y^\nu ,$$

associated with the Lorentz matrix

$$\begin{aligned} \Lambda^0{}_0 &= \frac{1}{\sqrt{1-\|v\|^2}} , & \Lambda^0{}_i &= -\frac{v_i}{\sqrt{1-\|v\|^2}} , \\ \Lambda^i{}_0 &= -\frac{v^i}{\sqrt{1-\|v\|^2}} , & \Lambda^i{}_j &= \delta^i_j + \frac{v^i v_j}{\sqrt{1-\|v\|^2}(1+\sqrt{1-\|v\|^2})} . \end{aligned} \tag{6.111}$$

If $(\bar{u}, \bar{r}, \bar{\theta}, \bar{\varphi})$ are Bondi coordinates associated with the barred Minkowskian coordinates $\bar{y}^\nu$, one easily checks that

$$\bar{u}\big|_{\bar{x}=0} = K u\big|_{x=0} , \tag{6.112}$$

with K as in Equation (6.45),

$$K(n) := \frac{\sqrt{1-\|v\|^2}}{1 - v_k n^k} = \frac{1 + v_k \bar{n}^k}{\sqrt{1-\|v\|^2}} , \qquad S^2 \ni n = (n^i) , \ \sum (n^i)^2 = 1 . \tag{6.113}$$

Let a cut S of $\mathscr{I}^+$ be given by the equation

$$S = \{u = 0\} \subset \mathscr{I}^+ ,$$

since $\Psi(\Lambda)$ is an isometry of b we have $\Psi(\Lambda)_* b = b$; further $\Psi(\Lambda)(S) = S$ by (6.111), and (6.109) reads

$$H(\Psi(\Lambda)_* X, S, g, b) = H(X, S, \Psi(\Lambda)^* g, b) . \tag{6.114}$$

Let, first $X = a$ be a translation,

$$a = a^\mu \partial_\mu ,$$

where the a^μ's are constants in the Minkowskian coordinates (6.110). Then

$$\bar{a} := \Psi(\Lambda)_* X = \Lambda^\mu{}_\nu a^\nu \partial_\mu ,$$

and Equation (6.114) together with the definition (6.18) of p^μ gives

$$\bar{p}^\mu \bar{a}_\mu = p^\mu a_\mu ,$$

which shows that p^μ transforms as a Lorentz vector under boosts, as desired.

It is instructive to obtain this result directly; from (C.127) and from the formulae for the behaviour of the Bondi functions under boosts of Appendix C.6 we have:

$$\begin{aligned}
\bar{p}^0 &= \frac{1}{4\pi}\int_{\bar{u}=0} \bar{M}\bar{\lambda} d^2x = \frac{1}{4\pi}\int_{u=0} \frac{\lambda M}{K} d^2x \\
&= \frac{1}{4\pi}\int_{u=0} M \frac{1 - v_k z^k}{\sqrt{1-\|v\|^2}} d^2x \\
&= \frac{p^0}{\sqrt{1-\|v\|^2}} - \frac{v_k p^k}{\sqrt{1-\|v\|^2}} , \\
\bar{p}^i &= \frac{1}{4\pi}\int_{\bar{u}=0} \bar{M}\ \bar{\lambda}\bar{z}^i d^2x \\
&= \frac{1}{4\pi}\int_{u=0} \lambda M \frac{\sqrt{1-\|v\|^2} z^i - v^i + \frac{v^i v_k z^k}{1+\sqrt{1-\|v\|^2}}}{\sqrt{1-\|v\|^2}} d^2x \\
&= p^i + \frac{v^i}{\sqrt{1-\|v\|^2}} \left(\frac{v_k p^k}{1+\sqrt{1-\|v\|^2}} - p^0 \right) .
\end{aligned}$$

Here, as elsewhere, $\lambda = \sqrt{\det h_{AB}}$. Consider, next, generators of the Lorentz group:

$$X(J) := J_\mu{}^\nu x^\mu \partial_\nu , \tag{6.115}$$

where — in Minkowskian coordinates — the $J_{\mu\nu}$'s form an anti-symmetric tensor with constant coefficients. In analogy to (6.92) we define

$$\frac{1}{2} L^{\mu\nu} J_{\mu\nu} := -H_L(X(J), S, g, b) . \tag{6.116}$$

An argument identical to the one above, using Equation (6.114), shows that the part of $L^{\mu\nu}$'s that arises from $H_L(X(J), S, g, b)$ transforms as a tensor in Minkowski space-time under Lorentz transformations. This does not suffice to control the behaviour of the Lorentz charges under boosts, because it is not clear what the transformation properties of the correction term in Equation (6.57) are. In order to show that things behave as expected, we use equations (6.60), (6.72) and (6.64) to rewrite (6.61) as

$$H_L(X,\mathscr{S}) = -\frac{1}{64\pi}\int_{S^2}\Big(24N_A + 2\chi_{AB}\chi^{BC}{}_{||C} + \frac{1}{2}(\chi_{BC}\chi^{BC})_{||A}\Big)\, X^A\big|_{x=0}\sin\theta\, d\theta\, d\varphi\,. \tag{6.117}$$

This formula holds both for boost and rotations. It follows from Equation (C.128) of Appendix C.6 that the term involving N_A is invariant under boosts:

$$\overline{\lambda}\,\overline{N}_A = \lambda\, N_A\,. \tag{6.118}$$

Next, it follows from the remaining formulae of Appendix C.6 that we have

$$\bar\lambda\left(2\bar\chi_{AB}\bar\chi^{BC}{}_{\bar{||}C} + \bar\chi^{BC}\bar\chi_{BC\bar{||}A}\right) = \lambda\left(2\chi_{AB}\chi^{BC}{}_{||C} + \chi^{BC}\chi_{BC||A}\right) + \lambda\left(2\chi_A{}^B\chi_B{}^C - \delta_A{}^C\chi_{BD}\chi^{BD}\right)(\ln K)_{||C}\,. \tag{6.119}$$

Since χ_{AB} is symmetric and traceless the expression in front of $(\ln K)_{||C}$ in (6.119) vanishes — this is easily seen by calculating this expression in a frame in which χ_{AB} is diagonal — and the Lorentz covariance of our definition of Lorentz charges follows.

We end this section with an explicit expression for the $L^{\mu\nu}$'s; setting

$$L_i := \frac{1}{2}\epsilon_{ijk}L^{jk} \iff L^{jk} = \epsilon^{jki}L_i$$

we are led to

$$L_i = \frac{1}{32\pi}\int_{S^2}\left(12N_A + \chi_{AB}\chi^{BC}{}_{||C}\right)\varepsilon^{AD}\partial_D n_i\,\sin\theta\, d\theta\, d\varphi\,, \tag{6.120}$$

$$L^{0i} = -\frac{1}{32\pi}\int_{S^2}\left\{\left(12N_A + \chi_{AB}\chi^{BC}{}_{||C}\right)(n^i)^{||A} - \frac{1}{2}\chi_{AB}\chi^{AB}n^i\right\}\sin\theta\, d\theta\, d\varphi\,. \tag{6.121}$$

In (6.121) the functions $n^i \in \{\sin\theta\cos\varphi, \sin\theta\sin\varphi, \cos\theta\}$ are understood as scalar functions on S^2.

6.9 BMS invariance of energy-momentum

In this section we will prove, for the sake of completeness, the well known result that the Trautman-Bondi energy-momentum vector is invariant under

BMS transformations; we follow the presentation of [43]. We shall consider cuts S of $\mathscr{I}$ which, in Bondi coordinates, are given by the equation

$$S = \{u = s(\theta, \phi)\} ,$$

for some, say smooth, function s on S^2. equations (C.100) and (C.99) give

$$\begin{aligned} H(X, \mathscr{S}^\epsilon) &= \int_{S^2} (\mathbb{W}^{xu} - \mathbb{W}^{xA}\partial_A s)\, d\theta\, d\varphi \\ &= \frac{1}{16\pi} \int_{S^2} \Big[(4M - \chi^{AB}{}_{||AB}) X^u + \frac{1}{2} \left(X^u \chi^{AB}{}_{||B} \right)_{||A} \\ &\quad + X^u \frac{1}{2} \dot\chi^{AB}{}_{||B} \partial_A s \Big] \sin\theta\, d\theta\, d\varphi\,. \end{aligned} \tag{6.122}$$

Similarly to (C.101) we verify that when $\partial_u X^u = 0$, then on the sphere $\{u = s\}$ we have

$$\int_{S^2 = \{u=s\}} \left[\left(X^u \chi^{AB}{}_{||B} \right)_{||A} + X^u \dot\chi^{AB}{}_{||B} \partial_A s \right] \sin\theta\, d\theta\, d\varphi = 0\,. \tag{6.123}$$

One is thus led to the following generalisation of Equation (6.18):

$$m := \frac{1}{16\pi} \int_{S^2} \left(4M - \chi^{AB}{}_{||AB}\right) (u = s(\theta, \phi), \theta, \phi) \sin\theta\, d\theta\, d\phi\,, \tag{6.124}$$

$$p^k := \frac{1}{16\pi} \int_{S^2} \left(4M - \chi^{AB}{}_{||AB}\right) (u = s(\theta, \phi), \theta, \phi) n^k \sin\theta\, d\theta\, d\phi\,, \tag{6.125}$$

where n^k, $k = 1, 2, 3$ denotes the functions $\sin\theta\cos\phi$, $\sin\theta\sin\phi$ and $\cos\theta$, in that order.

We recall the transformation laws for χ and M (*cf.*, *e.g.* [83]), which can also be found in the Appendix C.5:

$$\begin{aligned} \overline{M}(\bar u = u - \alpha(x^A), x^A) &= \Big[M + \frac{1}{2} \chi^{AB}{}_{,u||B} \alpha_{||A} + \frac{1}{4} \chi^{AB}{}_{,u} \alpha_{||AB} \\ &\quad + \frac{1}{4} \chi^{AB}{}_{,uu} \alpha_{||A} \alpha_{||B} \Big] (u, x^A)\,, \end{aligned}$$

$$\overline{\chi}_{AB}(\bar u = u - \alpha(x^C), x^C) = \left[\chi_{AB} - 2\alpha_{||AB} + \breve{h}_{AB} \Delta_2 \alpha \right] (u, x^C)\,.$$

Consider the quantity $\overline{\chi}^{AB}{}_{||\bar A \bar B}$, where $||\bar A \bar B$ denotes covariant derivatives with respect to the transformed coordinates, $\partial_{\bar A} = \partial_A + \alpha_{,A}\partial_0$. Now, the occurence of u derivatives in $\partial_{\bar A}$ will introduce u derivatives of χ_{AB} in the transformation formula for this quantity. With some work one finds that the combination $4M - \chi^{AB}{}_{||AB}$ has a simple transformation law with respect to the super-translations:

$$\overline{[4M - \chi^{AB}{}_{||AB}]}(\bar{u} = u - \alpha(\theta,\phi),\, \theta,\, \phi)$$
$$= \left[4M - \chi^{AB}{}_{||AB} + \Delta_2(\Delta_2 + 2)\alpha\right](u,\theta,\phi)\ .$$

The overbar in the left hand side of the last equation denotes the corresponding quantity calculated in the new Bondi frame. The inhomogeneous term involving α in this equation integrates out to zero when inserted into Equations (6.124)-(6.125): this is obvious for Equation (6.124), while in Equation (6.125) one has to use the fact that the functions n^i are in the kernel of the operator $\Delta_2 + 2$. This establishes invariance of the Trautman-Bondi four-momentum, under supertranslations understood in a passive way; by this we mean a change of background associated with a change of Bondi coordinates, with the cut of $\mathscr{I}$ under consideration remaining the same.

6.10 Polyhomogeneous Scri's

In this section we wish to extend part of our analysis to polyhomogeneous Scris, as considered in [44]. Let us therefore assume that the metric is polyhomogeneous, that is, all the functions appearing in the metric (5.93) have an asymptotic expansion of the form

$$f \approx \sum_i \sum_{j=1}^{N_i} f_{ij}(\theta,\varphi) x^i \ln^j x\ , \tag{6.126}$$

where the f_{ij}'s are smooth functions of their arguments. Here the expansion is meant in the sense that f can be approximated up to $O(r^{-N})$, for any N, by an appropriate finite number of terms from the right-hand side of (6.126). Further the same holds for any finite number of derivatives of f.

Suppose, first, that the free characteristic data h_{AB} have a polyhomogeneous expansion of the form

$$h_{AB} = \breve{h}_{AB} + x\chi_{AB} + x\ln x\, D_{AB} + O(x)\ , \tag{6.127}$$

so that

$$h^{AB} = \breve{h}^{AB} - x\chi^{AB} - x\ln x\, D^{AB} + O(x^2)\ . \tag{6.128}$$

As shown in [44], the vacuum Einstein equations imply then

$$x^{-2}U^A = O(\ln x)\ , \tag{6.129}$$

$$\beta = O(x^2(\ln x)^2)\ , \tag{6.130}$$

$$xV - 1 = O(x(\ln x)^2)\ , \tag{6.131}$$

$$\partial_u D_{AB} = 0\ .$$

With this asymptotic behaviour the integrand of (5.112) for $X = \partial_u$ contains terms which diverge in the limit as ϵ tends to zero. Somewhat surprisingly, it has been observed in [43] that those terms integrate out to zero, and formula (5.115) remains valid. However, with the asymptotic behaviour above, the boundary part of the "symplectic" form diverges:

$$16\pi \int_{\mathscr{S}_\epsilon} \delta p^0_{AB} \wedge \delta \mathfrak{g}^{AB} \approx \int_{\mathscr{S}_\epsilon} \frac{dx}{x} \sin\theta \, d\theta \, d\varphi \delta D_{AB} \wedge \delta \chi^{AB}$$
$$\approx \ln \epsilon \int_{S^2} \sin\theta \, d\theta \, d\varphi \delta D_{AB} \wedge \delta \chi^{AB} \, .$$

It follows that the occurrence of $x \ln x$ terms in h_{AB} is not compatible with a Hamiltonian framework.

Consider, next, polyhomogeneous metrics in which the logarithmic terms in h_{AB} start at order x^2:

$$h_{AB} = \breve{h}_{AB} + x\chi_{AB} + x^2 d_{AB}(\ln x) + O(x^2) \, , \qquad (6.132)$$

where $d_{AB}(\ln x)$ is a polynomial in $\ln x$ of order N, with coefficients which are smooth, symmetric, twice covariant tensor fields on S^2. The resulting asymptotic behaviour of the other Bondi functions, as well as of the momenta $p^\alpha{}_{\beta\gamma}$ can be found in Appendix C.1.2. Equations (C.31)–(C.44) show that the asymptotic behaviour of all the $p^\alpha{}_{\beta\gamma}$'s is the same as in the smooth case (5.98), modulo some multiplicative logarithmic factors which do not affect finiteness of all the volume integrals involved. Further, the asymptotic behaviour obtained preserves convergence of various boundary integrals; in particular, equations (5.104)–(5.109) remain true; the energy integral (5.115) converges as before (without any miraculous cancellations, as was the case with the asymptotic behaviour (6.127)); equations (5.116)–(5.118) again hold; the variational formula (5.119) remains valid with $X = \frac{\partial}{\partial u}$.

We note, nevertheless, that the asymptotic behaviour (6.132) leads to a potentially divergent term in the Freud potential (5.114),

$$-x^{-2} X^A e^{-2\beta} h_{AB} U^B_{,x} = x^{-1} X^A \chi^B{}_{A\|B} + 2X^A \breve{h}_{AB} W^B_{N+1} \ln^{N+1} x + \dots \, ,$$

for some coefficient W^B_{N+1}. This might lead to a logarithmic divergence for boosts and rotations in the relevant Hamiltonians, which then could cease to be well defined. A definite statement of this kind would require checking the detailed structure of the potentially divergent terms, which we haven't done. In any case, no diverging terms occur for Lorentz group generators if the $\ln x$ terms in h_{AB} start at the x^3 level or higher — this occurs *e.g.* for the "minimal sequences", compatible with non-vanishing Weyl tensor at Scri, considered in [44].

A. Odd forms (densities)

The purpose of this Appendix is to recall the notion of an *odd* differential n-form on a manifold M; we follow the very clear approach of [122]. Locally, in a vicinity of a point x_0, such a form may be defined as an equivalence class $[(\alpha_n, \mathcal{O})]$, where α_n is a differential n-form defined in a neighbourhood U and $\mathcal{O}$ is an orientation of U; the equivalence relation is given by:

$$(\alpha_n, \mathcal{O}) \sim (-\alpha_n, -\mathcal{O}) ,$$

where $-\mathcal{O}$ denotes the orientation opposite to $\mathcal{O}$. Using a partition of unity, we may define odd forms globally, even if the manifold is non-orientable.

Odd differential n-forms on an m-dimensional manifold can be described using antisymmetric contravariant tensor densities of rank $r = (m - n)$ (see [113]). Indeed, if $f^{i_1 \dots i_r}$ are components of such a tensor density with respect to a coordinate system (x^i), then we may assign to f an odd n-form defined by the representative $(\alpha_n, \mathcal{O})$, where $\mathcal{O}$ is the local orientation carried by $(x^1, \dots, x^m)$ and

$$\alpha_n := f^{i_1 \dots i_r} \left(\frac{\partial}{\partial x^{i_1}} \wedge \dots \wedge \frac{\partial}{\partial x^{i_r}} \right) \rfloor (dx^1 \wedge \dots \wedge dx^m) .$$

In particular, within this description scalar densities (*i.e.*, densities of rank $m - n = 0$) are odd forms of maximal rank, whereas vector densities are odd $(m - 1)$-forms.

Odd n-forms are designed to be integrated over *externally oriented* n-dimensional submanifolds. An exterior orientation of a submanifold Σ is an orientation of a bundle of tangent vectors transversal with respect to Σ. The integral of an odd form $\tilde{\alpha}_n = [(\alpha_n, \mathcal{O})]$ over a n-dimensional submanifold D with exterior orientation $\mathcal{O}_{\text{ext.}}$ is defined as follows:

$$\int_{(D, \mathcal{O}_{\text{ext.}})} \tilde{\alpha}_n := \int_{(D, \mathcal{O}_{\text{int.}})} \alpha_n ,$$

where $\mathcal{O}_{\text{int.}}$ is an internal orientation of D, such that $(\mathcal{O}_{\text{ext.}}, \mathcal{O}_{\text{int.}}) = \mathcal{O}$; it should be obvious that the result does not depend upon the choice of a representative. For example, a flow through a hypersurface depends usually upon its *exterior* orientation (given by a transversal vector) and does not feel any *interior* orientation. Similarly, the canonical formalism in field theory uses structures, which are defined in terms of flows through Cauchy hypersurfaces in space-time. This is why canonical momenta are described by odd $(m-1)$-forms. The integrals of such forms are insensitive to any *internal* orientation of the hypersurfaces they are integrated upon, but are sensitive to a choice of the time arrow (*i.e.*, to its *exterior* orientation).

The Stokes theorem generalizes to odd forms in a straightforward way:

$$\int_{(D,\mathcal{O}_{\text{ext}})} d\tilde{\alpha}_{n-1} = \int_{\partial(D,\mathcal{O}_{\text{ext}})} \tilde{\alpha}_{n-1} ,$$

where $d[(\alpha_n, \mathcal{O})] := [(d\alpha_n, \mathcal{O})]$ and $\partial(D, \mathcal{O}_{\text{ext}})$ is the boundary of D, equipped with an exterior orientation inherited in the canonical way from $(D, \mathcal{O}_{\text{ext}})$. This means that if $(e_1, \ldots, e_{m-n})$ is an oriented basis of vectors transversal to D and if f is a vector tangent to D, transversal to ∂D and pointing outwards of D, then the exterior orientation of $\partial(D, \mathcal{O}_{\text{ext}})$ is given by $(e_1, \ldots, e_{m-n}, f)$.

B. Solutions of the wave equation smoothly extendable to $\mathscr{I}^+$

The aim of this appendix is to discuss shortly the question of existence of solutions of the wave equation which are smooth on $\widetilde{M} \equiv M \cup \mathscr{I}^+$. From standard results in the theory of hyperbolic partial differential equations one has:

- A solution of Equation (4.22) is uniquely determined to the *future* of some hyperboloid $\mathscr{S}_\tau$ by prescribing $\tilde{f} = f/\Omega$ on $\mathscr{S}_\tau$ together with the derivative of $\tilde{f}$ in a direction transverse to $\mathscr{S}_\tau$. (Those data will be referred to as *standard* Cauchy data on $\mathscr{S}_\tau$.) In particular, prescribing $\tilde{f}$ and $\tilde{p}$ on $\mathscr{S}_\tau$ determines a unique solution of Equation (4.22) on the causal future $J^+(\mathscr{S}_\tau)$ of $\mathscr{S}_\tau$.
- A solution of Equation (4.22) is uniquely determined to the *past* of some hyperboloid $\mathscr{S}_\tau$ by prescribing standard initial data on $\mathscr{S}_\tau$ *and* prescribing $\tilde{f}$ on the portion of $\mathscr{I}^+$ lying to the past of $\mathscr{S}_\tau$ — that is, prescribing c on $\mathscr{I}^+_{(-\infty,\tau]}$.

Moreover, to any standard Cauchy data prescribed on $\mathscr{S}_\tau$ there corresponds a solution of Equation (4.22) on $J^+(\mathscr{S}_\tau)$. On the other hand, given standard Cauchy data on $\mathscr{S}_\tau$ and c on $\mathscr{I}^+_{(-\infty,\tau]}$ there will exist a solution of Equation (4.22), which is smoothly extendable to $\mathscr{I}^+$ if and only if certain obvious "corner conditions" are met. For example, the value of $\tilde{f}|_{S_\tau}$ as calculated from $\tilde{f}|_{\mathscr{S}_\tau}$ has to agree with that calculated from $c|_{S_\tau}$:

$$\tilde{f}\,|_{S_\tau} = c|_{\omega=\tau} \tag{B.1}$$

(otherwise the solution couldn't be continuous on $\widetilde{M}$). Next, from the standard Cauchy data on $\overline{\mathscr{S}}_\tau$ one can calculate all first derivatives of $\tilde{f}$ on $\overline{\mathscr{S}}_\tau$. Again this has to agree with those derivatives of $\tilde{f}$ which can be calculated from c if one wants $\tilde{f}$ to be differentiable on $\widetilde{M}$. This requirement, expressed in terms of the fields $(\tilde{f}, \tilde{p}, c)$, leads to the constraint

$$\left\{ \frac{\tilde{p}}{\sin\theta} - \frac{\partial \tilde{f}}{\partial \rho} \right\}\bigg|_{S_\tau} = \frac{\partial c}{\partial \omega}\bigg|_{\omega=\tau} \tag{B.2}$$

(see (4.25); we note that $\tilde{p}/\sin\theta$ is a smooth function on S^2 when $\tilde{p}$ is a smooth density). There is actually an infinite[1] hierarchy of conditions of that kind, which have to hold on S_τ if one wants to enforce smooth extendability of $\tilde{f}$ to $\mathscr{I}^-$ (in the sense that f/Ω can be extended by continuity to a smooth function on $\widetilde{M}$). To see this, recall that because $\overline{\mathscr{S}}_\tau$ is non-characteristic one can, using the standard Cauchy data and the wave Equation (4.1) (or actually its conformally rescaled version, Equation (4.22)), calculate all the partial derivatives of the solution on $\overline{\mathscr{S}}_\tau$, of any order, and in all directions. In particular, one can calculate all the derivatives of the solutions along $\mathscr{I}^+$. But those derivatives can also be calculated from c. To have a smooth solution those derivatives have to match. We emphasize that this is a relatively mild restriction: Borel's Lemma shows that for any standard Cauchy data on $\overline{\mathscr{S}}_\tau$ there will be a c_0 which satisfies those conditions. Then any c of the form $c_0 + c_1$, where c_1 is any function on $\mathscr{I}^+_{(-\infty,\tau]}$ which vanishes on S_τ together with all derivatives, will satisfy the corner conditions.

We note that while the solutions discussed above have the property that f/Ω extends by continuity to a smooth function defined on $M \cup \mathscr{I}^+$, there is no reason for the solutions so constructed to have any decent differentiability properties at i^0 or $\mathscr{I}^-$. In fact, for any standard initial data on $\mathscr{S}_\tau$ there will be "a lot of" c's on $\mathscr{I}^+_{(-\infty,\tau]}$ that will lead to a solution of Equation (4.22), which smoothly extends both across i^0 and $\mathscr{I}^-$: to obtain such c's one can extend the standard initial data in any way across S_τ to initial data for the wave equation on the Einstein cylinder (*cf.*, *e.g.*, [125, Fig. 11.1]) and solve the wave Equation (4.22) there. It follows from conformal covariance of Equation (4.22) that the solution f so obtained will indeed extend smoothly across i^0 and $\mathscr{I}^-$; one can then read off c from f. Different extensions of the standard initial data across S_τ will lead to different c's. However, it can be seen, using the so-called transport equations, that not all smooth c's on $\mathscr{I}^+_{(-\infty,\tau]}$ will be obtained in this way. It would be of some interest to give a complete description of those sets $(\tilde{f}, \tilde{p}, c)$ that lead to solutions f with good differentiability properties at i_0 and/or at $\mathscr{I}^-$. Another interesting problem is to give a complete description of those sets $(\tilde{f}, \tilde{p}, c)$ that lead to solutions f, which extend smoothly to $\mathscr{I}^+$ and which have finite energy on the hypersurfaces of standard Minkowskian time.

[1] If one is interested in extensions of f/Ω which are only C^k on $\widetilde{M}$, for some finite k, there will only be a finite number of corner conditions to impose.

C. Gravitational field: some auxiliary results

In this appendix we prove some auxiliary results, needed in the body of the paper.

C.1 The canonical gravitational variables in Bondi coordinates

Let, as elsewhere, $\breve{h}$ denote the standard round metric on S^2. Let the background connection $B^{\alpha}_{\sigma\mu}$ be the Levi-Civita connection of the flat metric

$$\begin{aligned} b = b_{\mu\nu}dx^{\mu}dx^{\nu} &\equiv -du^2 - 2du\,dr + r^2\breve{h}_{AB}dx^A dx^B \\ &= -du^2 + 2x^{-2}du\,dx + x^{-2}\breve{h}_{AB}dx^A dx^B \,. \end{aligned} \tag{C.1}$$

The non-vanishing $B^{\sigma}{}_{\beta\gamma}$'s are

$$B^u{}_{AB} = x^{-1}\breve{h}_{AB}, \quad B^x{}_{AB} = x\breve{h}_{AB}, \quad B^A{}_{xB} = -x^{-1}\delta^A_B, \tag{C.2}$$

$$B^x{}_{xx} = -2x^{-1}, \quad B^A{}_{BC} = \breve{\Gamma}^A{}_{BC}(\breve{h}) \,. \tag{C.3}$$

C.1.1 Smooth Scri's

Consider a metric of the Bondi form (5.93)–(5.94), with the asymptotic behaviour (5.98)–(5.103); we have $\sqrt{|\det g_{\mu\nu}|} = x^{-4}e^{2\beta}\sin\theta$, and the definition (5.3) leads to

$$16\pi\mathfrak{g}^{ux} = x^{-2}\sin\theta \,, \quad g^{ux} = x^2 e^{-2\beta} \,, \tag{C.4}$$

$$16\pi\mathfrak{g}^{AB} = x^{-2}e^{2\beta}\sin\theta h^{AB} \,, \quad g^{AB} = x^2 h^{AB} \,, \tag{C.5}$$

$$16\pi\mathfrak{g}^{xx} = xV\sin\theta \,, \quad g^{xx} = x^5 e^{-2\beta}V \,, \tag{C.6}$$

$$16\pi\mathfrak{g}^{xA} = x^{-2}U^A\sin\theta \,, \quad g^{xA} = x^2 e^{-2\beta}U^A \,, \tag{C.7}$$

$$\mathfrak{g}^{uA} = \mathfrak{g}^{uu} = g^{uA} = g^{uu} = 0 \,. \tag{C.8}$$

Further, (5.7) leads to the following expressions:

$$p^u{}_{xx} = p^u{}_{xA} = 0 \,, \tag{C.9}$$

$$p^u{}_{uu} = -\frac{e^{-2\beta}}{x} U_A U^A + \frac{1}{2} e^{-2\beta} (U_A U^A)_{,x} - V x^x \beta_{,x} \\ -\frac{x^2}{2} (xV - 1)_{,x} = M x^2 + O(x^3) , \tag{C.10}$$

$$p^u{}_{uA} = -\frac{1}{2} e^{-2\beta} x^2 (x^{-2} U_A)_{,x} = O(x^2) , \tag{C.11}$$

$$p^u{}_{ux} = \beta_{,x} = O(x) , \tag{C.12}$$

$$p^u{}_{AB} = \frac{1}{2} e^{-2\beta} x^2 (x^{-2} h_{AB})_{,x} - \frac{1}{2} x^2 (x^{-2} \breve{h}_{AB})_{,x} \\ = \frac{1}{2} e^{-2\beta} x^2 (x^{-2} h_{AB})_{,x} + x^{-1} \breve{h}_{AB} = -\frac{1}{2} \chi_{AB} + O(x) , \tag{C.13}$$

$$p^A{}_{xx} = 0 , \tag{C.14}$$

$$p^A{}_{xB} = -\frac{1}{2} h^{AC} h_{CB,x} + \delta^A{}_B \beta_{,x} = -\frac{1}{2} \chi^A{}_B + O(x) , \tag{C.15}$$

$$p^A{}_{uu} = -2U^A \partial_u \beta + h^{AB} \partial_u U_B - \frac{1}{2} e^{-2\beta} U^A \left(U^B U_B - V x^3 e^{2\beta} \right)_{,x} \\ -\frac{1}{2} \left(U^B U_B - V x^3 e^{2\beta} \right)^{||A} - \frac{U^A}{x e^{2\beta}} \left(U^B U_B - V x^3 e^{2\beta} \right) \\ = \frac{1}{2} \partial_u \chi^{AB}{}_{||B} x^2 + O(x^3) , \tag{C.16}$$

$$p^A{}_{uB} = \delta^A_B \partial_u \beta - U^A \beta_{,B} - \frac{1}{2} e^{-2\beta} U^A U_{B,x} - \frac{1}{2} (U_B{}^{||A} - U^A{}_{||B} + h^{AC} \partial_u h_{CB}) \\ + x^{-1} e^{-2\beta} U^A U_B = -\frac{1}{2} \partial_u \chi^A{}_B x + O(x^2) , \tag{C.17}$$

$$p^A{}_{ux} = -h^{AB} \left[e^{-2\beta} \beta_{,B} - \frac{1}{2} x^2 (x^{-2} U_B)_{,x} \right] = O(x^2) , \tag{C.18}$$

$$p^A{}_{BC} = \delta^A_B \beta_{,C} + \delta^A_C \beta_{,B} - \frac{U^A}{2x^2 e^{2\beta}} (x^2 h_{BC})_{,x} - \left(\Gamma^A{}_{BC}(h) - \breve{\Gamma}^A{}_{BC}(\breve{h}) \right) \\ = \frac{1}{2} x \left(\chi^{AD}{}_{||D} \breve{h}_{BC} + \chi_{BC}{}^{||A} - \chi^A{}_{B||C} - \chi^A{}_{C||B} \right) + O(x^2) , \tag{C.19}$$

$$p^x{}_{xx} = 0 , \tag{C.20}$$

$$p^x{}_{xA} = \frac{1}{2} e^{-2\beta} h_{AB} U^B{}_{,x} = -\frac{1}{2} x \chi^B{}_{A||B} + O(x^2) , \tag{C.21}$$

$$p^x{}_{uu} = -\frac{1}{2}\mathrm{e}^{-2\beta}\left[\partial_u - Vx^3\partial_x - U^A\partial_A\right]\left(U^BU_B - Vx^3\mathrm{e}^{2\beta}\right) - 2x^3V\partial_u\beta$$
$$+\mathrm{e}^{-2\beta}U^A\partial_uU_A - x^2V\mathrm{e}^{-2\beta}U^AU_A - x^3[1-(xV)^2] \tag{C.22}$$
$$= -\partial_u Mx^3 + O(x^4)\,,$$

$$p^x{}_{uA} = -\frac{1}{2}\mathrm{e}^{-2\beta}U^B(U_{A||B} + U_{B||A} + \partial_u h_{AB}) - \frac{1}{2}x^3(\mathrm{e}^{-2\beta}VU_{A,x} - V_{,A})$$
$$+U_Ax^2V\mathrm{e}^{-2\beta} = O(x^3)\,, \tag{C.23}$$

$$p^x{}_{ux} = -\frac{1}{2}\mathrm{e}^{-2\beta}U_AU^A_{\ ,x} + \frac{1}{2}x^2(xV)_{,x} + Vx^3\beta_{,x} + U^A\beta_{,A} + \partial_u\beta = O(x^2)\,, \tag{C.24}$$

$$p^x{}_{AB} = \frac{1}{2}\mathrm{e}^{-2\beta}\left(U_{A||B} + U_{B||A} + \partial_u h_{AB} + Vx^3h_{AB,x}\right) + x\breve{h}_{AB}$$
$$-Vx^2\mathrm{e}^{-2\beta}h_{AB} = \frac{1}{2}x\partial_u\chi_{AB} + O(x^2)\,. \tag{C.25}$$

C.1.2 Polyhomogeneous asymptotics

Consider a metric of the Bondi form (5.93)–(5.94), together with the following polyhomogeneous asymptotics:

$$h_{AB} = \breve{h}_{AB} + x\chi_{AB} + x^2\, d_{AB}(\ln x) + O(x^2)\,, \tag{C.26}$$

where $d_{AB}(\ln x)$ is a polynomial of order N in $\ln x$, with coefficients which are smooth tensor fields on S^2. It follows that

$$h^{AB} = \breve{h}^{AB} - x\chi^{AB} - x^2\, d^{AB}(\ln x) + O(x^2)\,. \tag{C.27}$$

It has been shown in [44] (see page 121 there) that one has

$$\partial_u d_{AB} = 0\,,$$

$$x^{-2}U^A = -\frac{1}{2}\chi^{AB}{}_{||B} + xW^A(\ln x) + O(x)\,, \tag{C.28}$$

where $W^A(\ln x)$ is a polynomial in $\ln x$ of order $N+1$ with coefficients which are smooth vector fields on S^2. Further,

$$\beta = -\frac{1}{32}x^2\chi^{AB}\chi_{AB} + O(x^3\ln^{2N} x)\,, \tag{C.29}$$

$$xV - 1 = -2Mx + O(x^2\ln^{N+1} x)\,. \tag{C.30}$$

The above inserted in Equations (C.9)–(C.25) leads to the following:

$$p^u{}_{xx} = p^u{}_{xA} = 0\,, \tag{C.31}$$

$$p^u{}_{uu} = O(x^2)\ , \tag{C.32}$$

$$p^u{}_{uA} = O(x^2 \ln^{N+1} x)\ , \tag{C.33}$$

$$p^u{}_{ux} = O(x)\ , \tag{C.34}$$

$$p^u{}_{AB} = -\frac{1}{2}\chi_{AB} + O(x \ln^{N-1} x)\ ,$$

$$p^A{}_{xx} = 0\ , \tag{C.35}$$

$$p^A{}_{xB} = -\frac{1}{2}\chi^A{}_B + O(x \ln^N x)\ , \tag{C.36}$$

$$p^A{}_{uu} = O(x^2)\ ,$$

$$p^A{}_{uB} = -\frac{1}{2}\partial_u \chi^A{}_B x + O(x^2)\ ,$$

$$p^A{}_{ux} = O(x^2 \ln^{N+1} x)\ , \tag{C.37}$$

$$p^A{}_{BC} = \frac{1}{2}x \left(\chi^{AD}{}_{||D}\breve{h}_{BC} + \chi_{BC}{}^{||A} - \chi^A{}_{B||C} - \chi^A{}_{C||B}\right) + O(x^2 \ln^N x)\ , \tag{C.38}$$

$$p^x{}_{xx} = 0\ , \tag{C.39}$$

$$p^x{}_{xA} = -\frac{1}{2}x\chi^B{}_{A||B} + O(x^2 \ln^{N+1} x)\ , \tag{C.40}$$

$$p^x{}_{uu} = O(x^3)\ , \tag{C.41}$$

$$p^x{}_{uA} = O(x^3)\ , \tag{C.42}$$

$$p^x{}_{ux} = O(x^2)\ , \tag{C.43}$$

$$p^x{}_{AB} = \frac{1}{2}x\partial_u \chi_{AB} + O(x^2)\ . \tag{C.44}$$

C.2 Solutions of the vacuum Einstein equations containing hyperboloidal hypersurfaces

In this section we wish to discuss the set of data needed to construct space-times which contain hyperboloidal hypersurfaces $\mathscr{S} = i(\Sigma)$ with compact interior, as defined in Section 5.6 (*cf.* Equation (5.95)), together with a "piece of $\mathscr{I}^+$" both to the future and to the past of $\mathscr{S}$. In order to do that, it is simplest to extend $\overline{\mathscr{S}}$ beyond its boundary to a larger three-dimensional manifold $\hat{\mathscr{S}}$ without boundary. Similarly, it is useful to extend the conformal Cauchy data $(\tilde{\gamma}, \tilde{K}, \omega, \dot{\omega})$ to fields $(\hat{\gamma}, \hat{K}, \hat{\omega}, \hat{\dot{\omega}})$ defined on $\hat{\mathscr{S}} = \hat{\imath}(\hat{\Sigma})$ (see the definition of $\mathscr{P}$ on page 98 for the meaning of those fields), where $\hat{\Sigma}$ is the corresponding extension of Σ and $\hat{\imath} : \hat{\Sigma} \to \hat{\mathscr{S}}$ is the associated extension of the diffeomorphism $i : \Sigma \to \mathscr{S} \subset M$. Recall that ω is a positive smooth up-to-boundary function on Σ that vanishes precisely on $\partial\Sigma$, while $\tilde{\gamma}$ is a Riemannian smooth up-to-boundary metric on Σ, which is a conformally rescaled equivalent of the metric γ induced by the physical space-time metric g on the hypersurface $\mathscr{S}$:

$$\tilde{\gamma} = \omega^2 \gamma , \qquad \gamma = i^* g .$$

Finally, $\tilde{K}$ is the pull-back to Σ of the extrinsic curvature of the hypersurface $\mathscr{S}$ in the conformal space-time $(\widetilde{M}, \tilde{g})$, with $\tilde{g} = \Omega^{-2} g$. We note that $\omega = \Omega \circ i$.

While the initial data (γ, K) have to satisfy the vacuum constraint equations, the extended fields will not be assumed to satisfy any constraints on $\hat{\Sigma} \setminus \Sigma$, thus we take any extension as differentiable as the original objects allow. The construction here will be local in a neighbourhood of $\hat{\mathscr{S}}$, and the resulting space-time $(\hat{M}, \hat{g})$ will be globally hyperbolic with Cauchy surface $\hat{\mathscr{S}}$. In this set-up $\mathscr{I}^+$ will arise as follows, *cf.* Figure C.1:

1. $\partial\mathcal{D}^+(\mathscr{S}; \hat{M}) \setminus \mathscr{S} = \mathscr{I}^+ \cap \partial J^+(\overline{\mathscr{S}}; \hat{M}) \setminus \partial\overline{\mathscr{S}}$;
2. $\partial J^-(\mathscr{S}; \hat{M}) \setminus \mathscr{S} = \mathscr{I}^+ \cap \partial J^-(\overline{\mathscr{S}}; \hat{M}) \setminus \partial\overline{\mathscr{S}}$.

Here and throughout ∂A denotes the topological boundary of a set A, $\partial A \equiv \overline{A} \setminus A$, while $J^+(A; B)$ denotes the causal future of a set A in a space-time B, $\mathcal{D}^+(A; B)$ denotes the future domain of dependence of a set A in a space-time B, *etc.* In plain words, the part of $\mathscr{I}^+$ which lies to the future of $\overline{\mathscr{S}}$ coincides with the boundary of the domain of dependence of $\mathscr{S}$ in the extended space-time $\hat{M}$, while the part of $\mathscr{I}^+$ which lies to the past of $\overline{\mathscr{S}}$ coincides with the boundary of the past of $\mathscr{S}$ in the extended space-time $\hat{M}$. This asymmetry in time, which arises from the fact that $\mathscr{I}^+$ is null in the case at hand, is the key to the understanding of the nature of the phase spaces which arise in the radiation regime. Now, the metric on $\mathcal{D}^+(\mathscr{S} = i(\Sigma); \hat{M})$ is uniquely determined by the initial data (γ, K) defined

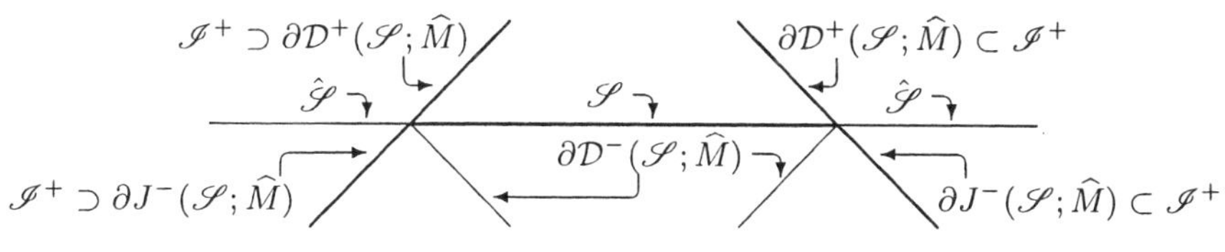

Fig. C.1. A hyperboloidal initial data surface $\mathscr{S}$ together with its extension $\hat{\mathscr{S}}$, in a conformally completed and extended space-time $\hat{M}$.

on Σ; by continuity those data determine uniquely all the relevant fields on $\partial\mathcal{D}^+(\mathscr{S};\widehat{M})$, and hence on that portion of $\mathscr{I}^+$ which lies to the future of $\mathscr{S}$. Those data also determine uniquely the metric on $\mathcal{D}^-(\mathscr{S};\widehat{M})$, but *not* on $J^-(\mathscr{S};\widehat{M})$. In order to obtain a unique metric on this last set one can proceed as follows: First, replacing $\widehat{M}$ by a subset thereof if necessary we can assume that $\partial\mathcal{D}^-(\mathscr{S};\widehat{M})$ and $\partial J^-(\overline{\mathscr{S}};\widehat{M})$ are smooth null hypersurfaces. Note that the metric on $\partial\mathcal{D}^-(\mathscr{S} = i(\Sigma);\widehat{M})$ and all its transverse derivatives are uniquely determined by the metric on $\mathcal{D}^-(\mathscr{S};\widehat{M})$, hence also by the "physical" initial data (Σ,γ,K). Now, $\partial\mathcal{D}^-(\mathscr{S};\widehat{M})$ and $\partial J^-(\overline{\mathscr{S}};\widehat{M})$ are two smooth null hypersurfaces that meet transversally at $\partial\mathscr{S} = \partial\overline{\mathscr{S}}$, and we can use the results of Friedrich [59, 60], as extended by Kánnár [89], concerning the asymptotic characteristic Cauchy problem for the gravitational field, to obtain a unique metric in $J^-(\overline{\mathscr{S}};\widehat{M})\setminus\mathcal{D}^-(\mathscr{S};\widehat{M})$: the knowledge of the metric and its transverse derivatives on $\partial\mathcal{D}^-(\mathscr{S};\widehat{M})$ determines uniquely the characteristic Cauchy data needed on $\partial\mathcal{D}^-(\mathscr{S};\widehat{M})$ in Kánnár's theorem [89], while (as shown in [43, Lemma III.1]) the fields $\chi_{AB}(u,\theta,\varphi)$ provide the data needed on $\partial J^-(\overline{\mathscr{S}};\widehat{M})$. Here we use a Bondi coordinate system on $\widetilde{M}$, defined in a neighbourhood of $\mathscr{I}^+\subset\widehat{M}$, to identify a tensor field $\chi_{AB}(\theta,\varphi)$ on "the abstract $\mathscr{I}^+$", discussed in Section 6.3, with a tensor field on $\mathscr{I}^+$. (There is a freedom involved in the choice of the Bondi coordinate system, we discuss this in some more detail below.) In fact, when appropriate initial data are given, Kánnár's theorem [89] gives existence and uniqueness of a vacuum space-time metric in a region forming a neighbourhood of $\partial\mathscr{S}$ in a manifold (with an "edge") with two boundaries meeting transversally at $\partial\mathscr{S}$, such that one of the boundaries is a subset of $\partial\mathcal{D}^-(\mathscr{S};\widehat{M})$, while the other corresponds to $\partial J^-(\overline{\mathscr{S}};\widehat{M})$.

The construction described so far provides a smooth metric on the union $\mathcal{D}^-(\mathscr{S};\widehat{M})\cup\mathcal{D}^+(\mathscr{S};\widehat{M})$, and a smooth metric on $J^-(\overline{\mathscr{S}};\widehat{M})\setminus\mathcal{D}^-(\mathscr{S};\widehat{M})$, such that the relevant Newman–Penrose functions (see [89] for details) providing the characteristic initial data match continuously across $\partial\mathcal{D}^-(\mathscr{S};\widehat{M})$. There is, however, no reason why transverse derivatives of the metric at $\partial\mathcal{D}^-(\mathscr{S};\widehat{M})$ should match smoothly across this hypersurface, and in general some of those derivatives will not. In order to obtain a smooth metric across $\partial\mathcal{D}^-(\mathscr{S};\widehat{M})$ appropriate corner conditions at $\partial\mathscr{S}$ involving (γ,K), χ_{AB}, and their derivatives at $\partial\mathscr{S}$ have to be satisfied. It is easy to understand the origin of those conditions: the vacuum Einstein equations, together with the initial data (γ,K) and a choice of gauges, determine uniquely all the transverse derivatives of the space-time metric on $\overline{\mathscr{S}}$. Knowing those derivatives allows one to calculate all the derivatives $\partial_u^i\chi_{AB}(u,\theta,\varphi)|_{u=0}$, $i\in\mathbb{N}$. However, for each choice of $\chi_{AB}(u,\theta,\varphi)$ satisfying the continuity requirement (C.53), discussed in detail below, one can obtain some smooth metric on $J^-(\overline{\mathscr{S}};\widehat{M})\setminus\mathcal{D}^-(\mathscr{S};\widehat{M})$. If $\chi_{AB}(u,\theta,\varphi)$ is such that its derivatives $\partial_u^i\chi_{AB}(u,\theta,\varphi)|_{u=0}$ do not match those calculated from (γ,K), then some of the derivatives of the space-time metric so obtained will not match across $\partial\mathcal{D}^-(\mathscr{S};\widehat{M})$. On the other hand, standard results in the theory of hyperbolic PDE's show that there exists $k \in \mathbb{N}$ such that if $k+n$ of the above described derivatives match at $\partial\mathscr{S}$, then the space-time metric will be n times continuously differentiable across $\partial\mathcal{D}^-(\mathscr{S};\widehat{M})$ (and hence everywhere if the initial data are smooth). In particular if all the derivatives match, then the space-time metric so constructed will be smooth. Given (γ, K), the requirement of the matching of a finite number of derivatives leads to algebraic equations for $\partial_u^i \chi_{AB}(u,\theta,\varphi)|_{u=0}$ which restrict the class of allowed associated $\chi_{AB}(u,\theta,\varphi)$'s. Similarly, the requirement of matching of all derivatives can be easily satisfied by a large class of χ_{AB}'s: Borel's summation Lemma shows that appropriate χ_{AB}'s exist, and one is free to prescribe $\chi_{AB}(u,\theta,\varphi)$ at will away from the matching surface $\partial\mathscr{S}$.

Let us examine in detail the corner conditions at $\partial\mathscr{S}$ that arise from the requirement of differentiability of the conformally rescaled space-time metric there. Consider, thus, a hyperboloidal initial data set (Σ,γ,K); we are interested in space-times for which $\mathscr{I}^+$ has $S^2\times\mathbb{R}$ topology, as in Minkowski space-time, and therefore we will assume

$$\partial\Sigma \approx S^2 \,, \tag{C.45}$$

where "$\approx$" means "is diffeomorphic to". Let $\tilde{h}$ be the metric induced by $\tilde{\gamma}$ on $\partial\Sigma$; since $\tilde{\gamma}$ is defined only up to a conformal factor, so is $\tilde{h}$, hence only the conformal class of $\tilde{h}$ is a geometrically well defined object. (We note that the discussion below is independent of the choice of the conformal factor involved.) Our hypothesis (C.45) together with the uniformization theorem show that there exist diffeomorphisms

$$\psi_\alpha : \partial\Sigma \to S^2$$

such that

$$\psi_\alpha^* \breve{h} = \phi_\alpha^2 \tilde{h} \,, \tag{C.46}$$

for some smooth strictly positive functions ϕ_α on $\partial\Sigma$, where the index α runs over the collection of all such maps; we will denote this collection by D. Clearly $\psi_\alpha \circ \psi_\beta^{-1} : S^2 \to S^2$ are conformal maps from $(S^2, \breve{h})$ to itself; any choice of ψ_β defines then a one-to-one correspondence between D and the conformal group G of $(S^2,\breve{h})$. Hence D can be thought of as the same set as the group G, in which the identity element has not been singled out. As is well known, G is isomorphic to the ortochronous Lorentz group (*cf.*, *e.g.*, [23, Vol. II, Thm. 18.10.4]), which shows that the maps ψ_α can be parameterized by elements of the Lorentz group whenever one such map ψ_β has been chosen.

Now, as already pointed out, the results of [61] show that there exists a smooth space-time $(\widehat{M},\hat{g})$ such that $\Omega^{-2}\hat{g}$ is vacuum in $\mathcal{D}(\mathscr{S};\widehat{M})$ for an appropriate positive conformal factor Ω, with Ω vanishing on $\partial\mathcal{D}^+(\mathscr{S};\widehat{M})$, so that this last set is a subset of $\mathscr{I}^+$ for the resulting vacuum space-time.

As shown in [118] in the smooth case, and in [44] in the polyhomogeneous case, for any choice of the map ψ_α in (C.46) one can find a Bondi coordinate system (u, r, x^A) as in Equation (5.93), with $u \in [0, u_1)$, $x \in [0, x_1)$, and with the collection of the domains of definition of the local coordinates (x^A) covering $\partial\mathscr{S}$, in which $\partial\mathscr{S}$ is given by the equation $x = u = 0$. Further, $\partial\mathcal{D}^+(\mathscr{S}; \widehat{M})$ will be given by the equation $x = 0$ in those coordinates, while $\overline{\mathscr{S}}$ will be given by an equation of the form

$$u = \alpha(x, x^A) \, , \qquad x \geq 0 \, ,$$

for some smooth function α, with $\alpha(0, x^A) = 0$. Let the conformal factor ω be defined as

$$\omega \equiv x \, . \tag{C.47}$$

The metric (5.93), with the coordinate r replaced by $1/x$, and conformally rescaled by r^{-2}, takes the form

$$\begin{aligned} \tilde{g}_{\mu\nu} \mathrm{d}x^\mu \mathrm{d}x^\nu &\equiv x^2 g_{\mu\nu} \mathrm{d}x^\mu \mathrm{d}x^\nu \\ &= -V x^3 \mathrm{e}^{2\beta} \mathrm{d}u^2 + 2\mathrm{e}^{2\beta} \mathrm{d}u \mathrm{d}x \\ &\quad + h_{AB} (\mathrm{d}x^A - U^A \mathrm{d}u)(\mathrm{d}x^B - U^B \mathrm{d}u) \, , \qquad (C.48) \\ \frac{\partial(\det h_{AB})}{\partial x} &= 0 \, . \qquad (C.49) \end{aligned}$$

We can use i to pull-back to Σ the coordinates (r, x^A); using the same symbols for the resulting coordinates on a neighbourhood, in Σ, of $\partial\Sigma \subset \overline{\Sigma}$, we obtain the following form of the pull-back to Σ of the metric $\tilde{g}$:

$$\begin{aligned} \tilde{\gamma} = &-V x^3 \mathrm{e}^{2\beta} \left(\alpha_{,x} \mathrm{d}x + \alpha_{,A} \mathrm{d}x^A\right)^2 + 2\mathrm{e}^{2\beta} \left(\alpha_{,x} \mathrm{d}x + \alpha_{,A} \mathrm{d}x^A\right) \mathrm{d}x \\ &+ h_{AB} \left(\mathrm{d}x^A - U^A \left(\alpha_{,x} \mathrm{d}x + \alpha_{,C} \mathrm{d}x^C\right)\right) \left(\mathrm{d}x^B - U^B \left(\alpha_{,x} \mathrm{d}x + \alpha_{,D} \mathrm{d}x^D\right)\right) \end{aligned} \tag{C.50}$$

We wish to show that χ_{AB} is, up to a multiplicative factor, the extrinsic curvature $\tilde{\lambda}_{AB}$ of $\partial\Sigma$ in $(\overline{\Sigma}, \tilde{\gamma})$. In order to see that, from the asymptotic behaviour of the various objects that occur in (C.50), Equations (5.98)–(5.100), we obtain the expansion

$$\tilde{\gamma} = (a^{-2} + O(x)) \mathrm{d}x^2 + O(x) \mathrm{d}x \mathrm{d}x^A + (\breve{h}_{AB} + x\chi_{AB} + O(x^2)) \mathrm{d}x^A \mathrm{d}x^B \, , \tag{C.51}$$

where

$$a^{-2} \equiv 2 \frac{\partial \alpha}{\partial x}\Big|_{x=0} \, .$$

We note that the length $|dx|_{\tilde{\gamma}}\Big|_{x=0}$ of dx with respect to the metric $\tilde{\gamma}$ equals a. Let $\tilde{n}$ be the $\tilde{\gamma}$–unit normal to $\partial\mathscr{S}$, we have

$$\tilde{n} = (a + O(x)) \partial_x + O(x) \partial_A \, , \tag{C.52}$$

with a similar expansion for the x^A-derivatives of $\tilde{n}$, hence

$$\begin{aligned}\tilde{\lambda}_{AB} &= -\frac{1}{2}\left(\tilde{n}^i\tilde{\gamma}_{AB,i} + \tilde{n}^i{}_{,A}\tilde{\gamma}_{iB} + \tilde{n}^i{}_{,B}\tilde{\gamma}_{Ai}\right)\Big|_{x=0} \\ &= -\frac{1}{2}a\chi_{AB} = -\frac{1}{2}|d\omega|_{\tilde{\gamma}}\chi_{AB}\ ,\end{aligned} \tag{C.53}$$

which is what we desired to show.

In order to be able to read off χ_{AB} from the initial data (Σ, γ, K) we need to determine $a = |d\omega|_{\tilde{\gamma}}|_{x=0} = |dx|_{\tilde{\gamma}}|_{x=0}$. The work required is done in Appendix C.3, where we show that

$$K|_{x=0} = -\frac{3}{\sqrt{2\alpha_{,x}}}\Big|_{x=0} = -3a\ . \tag{C.54}$$

This leads to

$$\chi_{AB} = \frac{6}{K}\tilde{\lambda}_{AB} \tag{C.55}$$

(recall that for hyperboloidal data sets K has no zeros on $\mathscr{I}$, *cf.*, *e.g.*, [4]). This is not quite the desired corner condition yet because of the following problem: in Equation (C.55) a conformal gauge has been used, in which the conformal factor ω equals the Bondi coordinate x; however, x is not known in terms of the initial data so far. In order to obtain an (approximate) formula for x, let $\tilde{\gamma}$ be any representative of the conformal class of $\tilde{\gamma}$, and let $(\overset{(0)}{x}, y^A)$ be a Gauss coordinate system for $\tilde{\gamma}$ near $\partial\mathscr{S}$, with $y^A = x^A$ at $\partial\mathscr{S}$ (recall that $x^A|_{\partial\mathscr{S}}$ has already been constructed in terms of initial data); we thus have

$$g = (\overset{(0)}{x}\phi)^{-2}\left(d(\overset{(0)}{x})^2 + \tilde{\gamma}_{AB}dy^A dy^B\right)\ ,$$

for some smooth up-to-boundary function ϕ without zeros. Set

$$\psi^{-2} \equiv \sqrt{\frac{\det\tilde{\gamma}_{AB}}{\det\breve{h}_{AB}}}\ , \qquad \overset{(1)}{x} \equiv \overset{(0)}{x}\phi\psi\ ; \tag{C.56}$$

there is a neighbourhood of $\partial\mathscr{S}$ on which $(\overset{(1)}{x}, y^A)$ can be used as coordinates. Choosing $(\overset{(1)}{x})^{-2}$ as the new conformal factor, replacing $\tilde{\gamma}$ by $(\overset{(1)}{x}/\overset{(0)}{x}\phi)^2\tilde{\gamma}$, and still denoting by $\tilde{\gamma}$ the resulting metric, we have

$$g = (\overset{(1)}{x})^{-2}\left(\tilde{\gamma}_{33}d(\overset{(1)}{x})^2 + O(\overset{(1)}{x})d\overset{(1)}{x}dy^A + \tilde{\gamma}_{AB}dy^A dy^B\right)\ , \tag{C.57}$$

with

$$\sqrt{\frac{\det\tilde{\gamma}_{AB}}{\det\breve{h}_{AB}}} = 1 + O((\overset{(1)}{x})^2)\ . \tag{C.58}$$

It follows from the above, and from the construction of Bondi coordinates on M as described *e.g.* in [44], that

$$\overset{(1)}{x} = x + O(x^2)\ , \qquad y^A = x^A + O(x)\ . \tag{C.59}$$

Equations (C.57) and (C.59) show that the extrinsic curvature $\tilde{\lambda}_{AB}$ of $\partial\mathscr{S}$ calculated in the conformal gauge (C.56)–(C.57) coincides with that calculated in the Bondi conformal gauge (C.47). Equation (C.55) allows one thus to relate the Bondi function χ_{AB} to the extrinsic curvature $\tilde{\lambda}_{AB}$ of $\partial\mathscr{S}$, calculated in any conformal gauge in which (C.57)–(C.59) hold.

From what has been said, and from the results described in [4], we obtain the following proposition:

Proposition C.2.1 *For any symmetric trace-free tensor field ψ_{AB} on S^2 there exists a vacuum space-time containing a hyperboloidal hypersurface $\mathscr{S}$ with compact interior such that the Bondi tensor field χ_{AB} at $\{u_0\} \times \partial\overline{\mathscr{S}}$ equals ψ_{AB}.*

This result is not quite satisfactory yet because one expects, on the basis of the work of Bondi *et al.* [25], as extended by Friedrich and Kánnár [59, 60, 89], that for any one-parameter family of symmetric trace-free tensor fields $\psi_{AB}(u)$ on S^2, $u \in [u_1, u_0]$, one should be able to find $\epsilon > 0$ and a vacuum space-time containing a hyperboloidal hypersurface $\mathscr{S}$ with compact interior such that the Bondi tensor field χ_{AB} on $(u_0 - \epsilon, u_0] \times \partial\overline{\mathscr{S}}$ equals $\psi_{AB}|_{(u_0-\epsilon,u_0]}$ (compare [43, Lemma III.1]). The proof of such a result would establish the uniqueness of the Trautman–Bondi mass in the class of monotonic Hamiltonians discussed in Section 5.10. In order to prove this, one would need to

1. work out the general form of the corner conditions at all orders, and
2. check that each of those conditions can be satisfied by an appropriate choice of hyperboloidal initial data (Σ, g, K). The problem here arises from the constraint equations, which have to be satisfied by the initial data. We expect that the results in [5, 6] can be used to show that no obstructions arise.

As a step towards the proof of such results, we note here the following form of the next corner condition, as worked out in Appendix C.3:

$$\partial_u \chi_{AB} = \left\{-4\tilde{n}\left[TS\left(xK_{AB}\right)\right] + \tilde{n}(K)\chi_{AB} - 2K^2 TS\left((K^{-2})_{\|AB}\right)\right\}\Big|_{x=0} \cdot$$

(In this formula we have assumed that α is constant on $\mathscr{I}^+$.)

C.3 Bondi coordinates *vs* hyperboloidal initial data

The object of this section is to derive formulae which allow one to read the Bondi functions from hyperboloidal initial data. While this has some interest of its own, it is also needed to obtain explicit — or recursive — equations for the corner conditions.

Consider, thus, a space-time metric of the Bondi form (5.93); if we introduce a new coordinate[1] ω by the formula

$$\omega := u - \alpha(x, \theta, \varphi) ,$$

then the metric induced on the level sets of ω has the following coordinate components:

$$\begin{aligned}\gamma_{xx} &= 2x^{-2}\mathrm{e}^{2\beta}\alpha_{,x} + \left(x^{-2}h_{AB}U^AU^B - xV\mathrm{e}^{2\beta}\right)\alpha_{,x}^2 \\ &= 2x^{-2}\alpha_{,x} + O(1) ,\end{aligned}$$

$$\begin{aligned}\gamma_{xA} &= x^{-2}\mathrm{e}^{2\beta}\alpha_{,A} - x^{-2}h_{AB}U^B\alpha_{,x} + \left(x^{-2}h_{CB}U^CU^B - xV\mathrm{e}^{2\beta}\right)\alpha_{,x}\alpha_{,A} \\ &= x^{-2}\alpha_{,A} + O(1) ,\end{aligned}$$

$$\begin{aligned}\gamma_{AB} = &\left(x^{-2}h_{CD}U^CU^D - xV\mathrm{e}^{2\beta}\right)\alpha_{,A}\alpha_{,B} - x^{-2}(\alpha_{,A}h_{BC}U^C + \alpha_{,B}h_{AC}U^C) \\ &+x^{-2}h_{AB} = x^{-2}\breve{h}_{AB} + x^{-1}\chi_{AB} + O(1) .\end{aligned}$$

We emphasize that the function α is uniquely determined by the choice of a Bondi coordinate system and a hyperboloidal initial data set, but this relation is highly non-trivial, and the first aim of the calculations presented here is to obtain equations which allow us to determine an asymptotic expansion for α; once this is done, the Bondi functions and the corner conditions will follow. To calculate the extrinsic curvature tensor of the level sets of ω,

$$K_{kl} = -N\Gamma^{\omega}_{kl} , \tag{C.60}$$

we need to calculate the lapse function $N \equiv 1/\sqrt{-g(\nabla\omega, \nabla\omega)}$ as well as the relevant Christoffel symbols; one has

$$\begin{aligned}N^{-2} &= 2x^2\mathrm{e}^{-2\beta}\alpha_{,x} - x^2h^{AB}\alpha_{,A}\alpha_{,B} - Vx^5\mathrm{e}^{-2\beta}\alpha_{,x}^2 - 2x^2\mathrm{e}^{-2\beta}U^A\alpha_{,A}\alpha_{,x} \\ &= 2x^2\alpha_{,x} - x^2\breve{h}^{AB}\alpha_{,A}\alpha_{,B} + x^3\chi^{AB}\alpha_{,A}\alpha_{,B} + O(x^4) ,\end{aligned} \tag{C.61}$$

and we note that spacelikeness of $\mathscr{S}$ is equivalent to $2\alpha_{,x} - \breve{h}^{AB}\alpha_{,A}\alpha_{,B} > 0$. The transformation rules for the Christoffel symbols give

$$\begin{aligned}\Gamma^{\omega}_{AB} = &\Gamma^u_{AB} - \alpha_{,k}\Gamma^k_{AB} + \alpha_{,AB} + \alpha_{,A}\alpha_{,B}\left(\Gamma^u_{uu} - \alpha_{,k}\Gamma^k_{uu}\right) \\ &+ \left(\Gamma^u_{uA} - \alpha_{,k}\Gamma^k_{uA}\right)\alpha_{,B} + \left(\Gamma^u_{uB} - \alpha_{,k}\Gamma^k_{uB}\right)\alpha_{,A} ,\end{aligned} \tag{C.62}$$

$$\begin{aligned}\Gamma^{\omega}_{xB} = &\Gamma^u_{xB} - \alpha_{,k}\Gamma^k_{xB} + \alpha_{,xB} + \alpha_{,x}\alpha_{,B}\left(\Gamma^u_{uu} - \alpha_{,k}\Gamma^k_{uu}\right) \\ &+ \left(\Gamma^u_{ux} - \alpha_{,k}\Gamma^k_{ux}\right)\alpha_{,B} + \left(\Gamma^u_{uB} - \alpha_{,k}\Gamma^k_{uB}\right)\alpha_{,x} ,\end{aligned} \tag{C.63}$$

[1] We hope that the conflict of notation between the coordinate ω here, and the three-dimensional conformal factor denoted by ω elsewhere in this paper, will not confuse the reader.

$$\begin{aligned} \Gamma^{\omega}_{xx} &= \Gamma^{u}_{xx} - \alpha_{,k}\Gamma^{k}_{xx} + \alpha_{,xx} + \alpha^2_{,x}\left(\Gamma^{u}_{uu} - \alpha_{,k}\Gamma^{k}_{uu}\right) \\ &\quad +2\left(\Gamma^{u}_{ux} - \alpha_{,k}\Gamma^{k}_{ux}\right)\alpha_{,x} \,. \end{aligned} \tag{C.64}$$

To calculate those objects, we recall Equation (5.7),

$$\Gamma^{\alpha}_{\mu\nu} = B^{\alpha}_{\mu\nu} - p^{\alpha}_{\mu\nu} + \delta^{\alpha}_{(\mu}\Gamma^{\sigma}_{\nu)\sigma} - \delta^{\alpha}_{(\mu}B^{\sigma}_{\nu)\sigma} \,.$$

From the identity $\Gamma^{\sigma}_{\rho\sigma} = \partial_\rho \ln(\det g_{\alpha\beta})/2$ one finds

$$\Gamma^{\sigma}_{\alpha\sigma} - B^{\sigma}_{\alpha\sigma} = 2\partial_\alpha\beta \,,$$

and using the formulae for $B^{\alpha}_{\beta\gamma}$ and $p^{\alpha}_{\beta\gamma}$ of Appendix C.1 we obtain

$$\Gamma^{u}_{uu} - \alpha_{,k}\Gamma^{k}_{uu} = -p^{u}_{uu} + \alpha_{,k}p^{k}_{uu} + 2\partial_u\beta = O(x^2) \,, \tag{C.65}$$

$$\Gamma^{u}_{uA} - \alpha_{,k}\Gamma^{k}_{uA} = -p^{u}_{uA} + \alpha_{,k}p^{k}_{uA} - \alpha_{,A}\partial_u\beta + \partial_A\beta = -\frac{1}{2}x\alpha_{,B}\partial_u\chi^{B}{}_{A} + O(x^2) \,, \tag{C.66}$$

$$\Gamma^{u}_{ux} - \alpha_{,k}\Gamma^{k}_{ux} = -p^{u}_{ux} + \alpha_{,k}p^{k}_{ux} + \beta_{,x} = O(x^2) \,, \tag{C.67}$$

$$\Gamma^{u}_{xx} = -p^{u}_{xx} = 0 \,, \qquad \Gamma^{u}_{xA} = -p^{u}_{xA} = 0 \,, \tag{C.68}$$

$$\begin{aligned} \Gamma^{u}_{AB} &= B^{u}_{AB} - p^{u}_{AB} \\ &= -\frac{1}{2}e^{-2\beta}x^2(x^{-2}h_{AB})_{,x} = x^{-1}\breve{h}_{AB} + \frac{1}{2}\chi_{AB} + O(x^2) \,, \end{aligned} \tag{C.69}$$

$$\begin{aligned} \Gamma^{x}_{AB} &= B^{x}_{AB} - p^{x}_{AB} \\ &= -\frac{1}{2}e^{-2\beta}\left(U_{A\|B} + U_{B\|A} + \partial_u h_{AB} + Vx^3 h_{AB,x} - 2Vx^2 h_{AB}\right) \\ &= x\breve{h}_{AB} - \frac{1}{2}x\partial_u\chi_{AB} + O(x^2) \,, \end{aligned} \tag{C.70}$$

$$\begin{aligned} \Gamma^{A}_{BC} &= B^{A}_{BC} - p^{A}_{BC} + 2\delta^{A}_{(B}\beta_{,C)} \\ &= \frac{1}{2}e^{-2\beta}x^{-2}U^{A}(x^2h_{BC})_{,x} + \Gamma^{A}_{BC}(h) \\ &= \breve{\Gamma}^{A}_{BC} - \frac{1}{2}x\left(\chi^{AD}{}_{\|D}\breve{h}_{BC} + \chi_{BC}{}^{\|A} - \chi^{A}{}_{B\|C} - \chi^{A}{}_{C\|B}\right) + O(x^2) \,, \end{aligned} \tag{C.71}$$

$$\Gamma^{x}_{xx} = B^{x}_{xx} - p^{x}_{xx} + 2\beta_{,x} = -2x^{-1} + 2\beta_{,x} \,, \tag{C.72}$$

$$\begin{aligned} \Gamma^{x}_{xA} &= B^{x}_{xA} - p^{x}_{xA} + \beta_{,A} = -\frac{1}{2}e^{-2\beta}h_{AB}U^{B}{}_{,x} + \beta_{,A} \\ &= \frac{1}{2}x\chi^{B}{}_{A\|B} + O(x^2) \,, \end{aligned} \tag{C.73}$$

$$\Gamma^A_{xx} = B^A_{xx} - p^A_{xx} = 0 \,, \tag{C.74}$$

$$\begin{aligned} \Gamma^A_{xB} &= B^A_{xB} - p^A_{xB} + \delta^A{}_B \beta_{,x} = -x^{-1}\delta^A_B + \frac{1}{2} h^{AC} h_{CB,x} \\ &= -x^{-1}\delta^A_B + \frac{1}{2}\chi^A{}_B + O(x^2) \,. \end{aligned} \tag{C.75}$$

Inserting the above in (C.62)–(C.64), one arrives at

$$\begin{aligned} \Gamma^\omega_{AB} &= x^{-1}\breve{h}_{AB} + \frac{1}{2}\chi_{AB} + \alpha_{\|AB} - x\alpha_{,x}(\breve{h}_{AB} - \frac{1}{2}\partial_u \chi_{AB}) \\ &+ \frac{1}{2} x\alpha_{,C}\left(\chi^{CD}{}_{\|D}\breve{h}_{AB} + \chi_{AB}{}^{\|C} - \chi^C{}_{B\|A} - \chi^C{}_{A\|B}\right) \\ &- \frac{1}{2} x\alpha_{,C}\left(\partial_u \chi^C{}_A \alpha_{,B} + \partial_u \chi^C{}_B \alpha_{,A}\right) + O(x^2) \,, \end{aligned} \tag{C.76}$$

$$\Gamma^\omega_{xx} = 2x^{-1}\alpha_{,x} + \alpha_{,xx} - 2\alpha_{,x}\beta_{,x} + O(x^2) \,, \tag{C.77}$$

$$\Gamma^\omega_{xA} = x^{-1}\alpha_{,A} + \alpha_{,xA} - \frac{1}{2}\alpha_{,B}\chi^B{}_A - \frac{1}{2}x\alpha_{,x}\left(\chi^B{}_{A\|B} + \partial_u \chi^B{}_A \alpha_{,B}\right) + O(x^2) \,. \tag{C.78}$$

We have the following implicit formulae for the inverse of the metric induced on $\mathscr{S}$:

$$\frac{1}{\gamma^{xx}} = x^{-2}\left[2\alpha_{,x} - \breve{h}^{AB}\alpha_{,A}\alpha_{,B} + x\chi^{AB}\alpha_{,A}\alpha_{,B} + O(x^2)\right] \,, \tag{C.79}$$

$$-\frac{\gamma^{xA}}{\gamma^{xx}} = \alpha^{,A} - x\chi^{AB}\alpha_{,B} + O(x^2) \,, \tag{C.80}$$

$$\gamma^{AB} = {}^2\gamma^{AB} + \frac{\gamma^{xA}\gamma^{xB}}{\gamma^{xx}} \,, \tag{C.81}$$

where ${}^2\gamma^{AC}\gamma_{CB} = \delta^A{}_B$, so that

$${}^2\gamma^{AB} = x^2\left[\breve{h}^{AB} - x\chi^{AB} + O(x^2)\right] \,.$$

In order to calculate the trace of the intrinsic curvature tensor of the level sets of ω, from (C.79)–(C.81) we obtain

$$-\frac{\gamma^{kl}K_{kl}}{N\gamma^{xx}} = \Gamma^\omega_{xx} + 2\frac{\gamma^{xB}}{\gamma^{xx}}\Gamma^\omega_{xB} + \frac{\gamma^{xA}}{\gamma^{xx}}\frac{\gamma^{xB}}{\gamma^{xx}}\Gamma^\omega_{AB} + \frac{{}^2\gamma^{AB}}{\gamma^{xx}}\Gamma^\omega_{AB} \,. \tag{C.82}$$

In what follows we will assume that $\alpha_{,A} = O(x)$ — such Bondi coordinates can always be constructed. It follows from Equations (C.60)–(C.64) and (C.82) that we can explicitly calculate all the derivatives of α at $x=0$ in terms of the initial data. For example, Equations (C.61), (C.82) and (C.60) lead to

$$N = \frac{1}{x\sqrt{2\alpha_{,x}+O(x^2)}}\,,$$

$$K \equiv \gamma^{kl}K_{kl} = -\frac{1}{\sqrt{2\alpha_{,x}}}\left[3 + x\frac{\alpha_{,xx}}{2\alpha_{,x}} + O(x^2)\right]\,. \tag{C.83}$$

This shows that

$$\alpha_{,x}\Big|_{x=0} = \frac{9}{2K^2}\,, \tag{C.84}$$

$$\alpha_{xx}\Big|_{x=0} = -\frac{1}{2}\left(\frac{3}{K}\right)^3 K_{,x}\,. \tag{C.85}$$

Let, as in Appendix C.2, $\tilde{n}$ be the field of $\tilde{\gamma}$-unit normals to $\partial\mathscr{S}$, from Equation (C.52) we thus obtain

$$\tilde{n}(\alpha) = -\frac{3}{2K}\Big|_{x=0}\,,$$

which gives a somewhat more geometric form of Equation (C.84). We note the formula, which follows from Equation (C.60),

$$K_{AB} = -\frac{1}{x^2\sqrt{2\alpha_{,x}}}\left(\breve{h}_{AB} + \frac{1}{2}x\chi_{AB} + O(x^2)\right)\,, \tag{C.86}$$

and which shows that χ_{AB} can also be read-off from K_{AB}. (This equation can be used to obtain an alternative form of the first order corner condition; one can use the results in [5] to check that the resulting equation is indeed consistent with (C.55).) To obtain the next order corner condition one needs a higher order version of Equation (C.86):

$$K_{AB} = -N\left\{\frac{1-x^2\alpha_{,x}}{x}\breve{h}_{AB} + \frac{1}{2}\chi_{AB} + \alpha_{\|AB} + \frac{1}{2}x\alpha_{,x}\partial_u\chi_{AB} + O(x^2)\right\}\,.$$

Let $TS(t_{AB}) := t_{AB} - \frac{1}{2}\breve{h}^{CD}t_{CD}\breve{h}_{AB}$ be the trace free part of t_{AB}; we then have

$$xTS(K_{AB}) = -\frac{1}{\sqrt{2\alpha_{,x}}}TS\left\{\frac{1}{2}\chi_{AB} + \alpha_{\|AB} + \frac{x}{2}\alpha_{,x}\partial_u\chi_{AB} + O(x^2)\right\}\,. \tag{C.87}$$

Calculating $x\alpha_{,x}\partial_u\chi_{AB}$ from (C.87) and differentiating in x, one obtains

$$\begin{aligned}\partial_u\chi_{AB} &= -\frac{2}{\alpha_{,x}}\left\{\partial_x\left[\sqrt{2\alpha_{,x}}xTS(K_{AB})\right] + TS(\alpha_{,x\|AB})\right\}\Big|_{x=0}\\ &= \left\{\frac{4K}{3}\partial_x\left[TS\left(xK_{AB}\right)\right] - \frac{K}{3}K_{,x}\chi_{AB} - 2K^2TS\left((K^{-2})_{\|AB}\right)\right\}\Big|_{x=0}\end{aligned}$$

$$= \left\{-4\tilde{n}\left[TS\left(xK_{AB}\right)\right] + \tilde{n}(K)\chi_{AB} - 2K^2 TS\left((K^{-2})_{\|AB}\right)\right\}\Big|_{x=0} \ . \quad \text{(C.88)}$$

Here we have used (C.84) and (C.85) to eliminate $\alpha_{,x}$ and $\alpha_{,xx}$. This gives the desired corner condition, when the right-hand side of (C.88) is calculated in any coordinate system (x, y^A) which agrees with the Bondi coordinate system on $\mathscr{S}$ one order higher than in (C.59). One can obtain one such coordinate system $(\overset{(2)}{x}, y^A)$ starting from the form (C.57) of the metric g: one needs, first, to readjust the y^A's of (C.57) by setting

$$y^A \to y^A + x\breve{h}^{AB}(\alpha_{,B} - \tilde{\gamma}_{3B}) \ .$$

This preserves (C.57)–(C.59). Next, one sets

$$\psi^{-2} \equiv \sqrt{\frac{\det \tilde{\gamma}_{AB}}{\det \breve{h}_{AB}}} = 1 + O((\overset{(1)}{x})^2) \ , \qquad \overset{(2)}{x} \equiv \overset{(1)}{x}\psi \ . \quad \text{(C.89)}$$

A conformal rescaling of $\tilde{\gamma}$ leads then to

$$g = (\overset{(2)}{x})^{-2}\left(\tilde{\gamma}_{33}d(\overset{(2)}{x})^2 + 2[\alpha_{,A} + O((\overset{(2)}{x})^2)]d\overset{(2)}{x}dy^A + \tilde{\gamma}_{AB}dy^A dy^B\right) \ , \quad \text{(C.90)}$$

with

$$\sqrt{\frac{\det \tilde{\gamma}_{AB}}{\det \breve{h}_{AB}}} = 1 + O((\overset{(2)}{x})^4) \ . \quad \text{(C.91)}$$

It can be checked that we have the desired property

$$\overset{(2)}{x} = x + O(x^3) \ , \qquad y^A = x^A + O(x^2) \ . \quad \text{(C.92)}$$

Proceeding as above, one can iteratively construct coordinates which agree with the Bondi ones to infinite order, using Borel's summation. The higher order corner conditions can be obtained by u differentiating the formula for K_{AB} that follows from (C.60) and (C.62), and by using the vacuum Einstein equations to recursively replace time derivatives of the various fields by space-derivatives of the initial data.

C.4 The calculation of the $\mathbb{W}^{x\mu}$'s

In this Appendix we shall present the details of the calculation of $\mathbb{W}^{xA}$, for any X satisfying the conditions that $X^u, x^{-1}X^x, X^A$ extend continuously to $\mathscr{I}^+$. From the definition (5.17) we have

$$\begin{aligned}\mathbb{W}^{xA} = {} & X^u\left[\mathfrak{g}^{\mu x}p^A{}_{\mu u} - \mathfrak{g}^{\mu A}p^x{}_{\mu u}\right] - \mathfrak{g}^{\alpha x}X^A{}_{;\alpha} + \mathfrak{g}^{\alpha A}X^x{}_{;\alpha} \\ & + X^x\left[\mathfrak{g}^{\mu x}p^A{}_{\mu x} - \mathfrak{g}^{\mu A}p^x{}_{\mu x} - \mathfrak{g}^{\mu\nu}p^A{}_{\mu\nu} + \frac{1}{3}\mathfrak{g}^{\mu A}p^\sigma{}_{\mu\sigma}\right] \\ & + X^B\left[\mathfrak{g}^{\mu x}p^A{}_{\mu B} - \mathfrak{g}^{\mu A}p^x{}_{\mu B} + (\mathfrak{g}^{\mu\nu}p^x{}_{\mu\nu} - \frac{1}{3}\mathfrak{g}^{\mu x}p^\sigma{}_{\mu\sigma})\delta^A_B\right] \ .\end{aligned} \quad \text{(C.93)}$$

Equations (5.98)–(5.100) give the following:

$$x^{-2}U^A = -\frac{1}{2}\chi^{AB}{}_{\|B} + x\left(2N_A + \frac{1}{2}\chi_{AB}\chi^{BC}{}_{\|C} + \frac{1}{16}(\chi_{CD}\chi^{CD})_{\|A}\right) + O(x^2)\,, \tag{C.94}$$

$$h_{AB} = \breve{h}_{AB} + x\chi_{AB} + \frac{1}{2}x^2\chi_A{}^C\chi_{CB} + O(x^3)\,, \tag{C.95}$$

$$h^{AB} = \breve{h}^{AB} - x\chi^{AB} + \frac{1}{2}x^2\chi^{AC}\chi_C{}^B + O(x^3)\,, \tag{C.96}$$

$$\beta = -\frac{1}{32}x^2\chi^{AB}\chi_{AB} + O(x^3)\,, \tag{C.97}$$

$$V = x^{-1} - 2M + O(x)\,. \tag{C.98}$$

Somewhat tedious calculations, using those equations together with formulae in Appendix C.1, lead to

$$\mathfrak{g}^{\mu x}p^A{}_{\mu u} - \mathfrak{g}^{\mu A}p^x{}_{\mu u} = \frac{\sin\theta}{16\pi}x^{-2}\dot{U}^A + O(x)\,,$$

$$\mathfrak{g}^{\mu x}p^A{}_{\mu x} - \mathfrak{g}^{\mu A}p^x{}_{\mu x} - \mathfrak{g}^{\mu\nu}p^A{}_{\mu\nu} + \frac{1}{3}\mathfrak{g}^{\mu A}p^\sigma{}_{\mu\sigma} =$$

$$-\frac{\sin\theta}{x16\pi}\left(\frac{1}{2}U^A{}_{,x} - \chi^{AB}{}_{\|B}\right) + O(1)\,,$$

$$\mathfrak{g}^{\mu x}p^A{}_{\mu B} - \mathfrak{g}^{\mu A}p^x{}_{\mu B} + (\mathfrak{g}^{\mu\nu}p^x{}_{\mu\nu} - \frac{1}{3}\mathfrak{g}^{\mu x}p^\sigma{}_{\mu\sigma})\delta^A_B =$$

$$-\frac{\sin\theta}{16\pi}\left[\underline{x^{-2}h^{AC}\dot{h}_{CB}} + x^{-2}U_B{}^{\|A} - 2x^{-2}\dot{\beta} - x^{-2}U^C{}_{\|C}\right] + O(x)\,.$$

We note that the underlined term in this last equation gives a potentially divergent contribution to the integrals involving boost, as well as rotational, background Killing vectors. Further,

$$\Delta\mathfrak{g}^{\alpha x}X^A{}_{;\alpha} - \Delta\mathfrak{g}^{\alpha A}X^x{}_{;\alpha} =$$

$$\frac{\sin\theta}{16\pi}\left(2MX^A - \frac{1}{2}X^A{}_{\|B}\chi^{BC}{}_{\|C} + \chi^{AB}X_B + x^{-1}X^x{}_{\|B}\chi^{AB}\right) + O(x)\,.$$

It follows that

$$\mathbb{W}^{Ax} - \mathbb{W}^{Ax}\Big|_{g=b} = \frac{\sin\theta}{16\pi}\Big\{X^u\frac{1}{2}\dot{\chi}^{AB}{}_{\|B} - x^{-1}X^x\frac{3}{2}\chi^{AB}{}_{\|B} + \tag{C.99}$$

$$+X^B\left[\underline{x^{-1}\dot{\chi}^A{}_B} + \delta^A_B\left(\frac{1}{8}\chi_{CD}\dot{\chi}^{CD} + \frac{1}{2}\chi^{CD}{}_{\|CD}\right) - \frac{1}{2}\chi_B{}^C{}_{\|C}{}^A - \chi^{AC}\dot{\chi}_{CB}\right]$$

$$+\left[2MX^A - \frac{1}{2}X^A_{\|B}\chi^{BC}{}_{\|C} + \chi^{AB}X_B + x^{-1}X^x{}_{\|B}\chi^{AB}\right]\Big\} + O(x)\,.$$

To obtain Equations (6.14) and (6.19), we note that it follows from Equations (C.94)–(C.98) that some of the relevant terms in (5.114) have the following behaviour

$$2x^{-2} - x^{-2}e^{2\beta}h^{AB}\check{h}_{AB} = -\frac{3}{8}\chi_{AB}\chi^{AB} + O(x)\,,$$

$$x^{-2}e^{-2\beta}h_{AB}U^B{}_{,x} = -x^{-1}\chi_A{}^B{}_{\|B} + 6N_A + \frac{1}{2}\chi_{AB}\chi^{BC}{}_{\|C} + \frac{3}{16}(\chi_{CD}\chi^{CD})_{\|A} + O(x)\,,$$

which gives

$$\begin{aligned}\mathbb{W}^{xu}\Big|_{\mathscr{I}^+} - \mathbb{W}^{xu}\Big|_{g=b} = \frac{\sin\theta}{16\pi}\Big[&(4M - \chi^{AB}{}_{\|AB})X^u + \frac{1}{2}\left(X^u\chi^{AB}{}_{\|B}\right)_{\|A}\\ &+\underline{x^{-1}X^A\chi_A{}^B{}_{\|B}} - \frac{3}{8}x^{-1}X^x\chi_{AB}\chi^{AB}\\ &-\left(6N_A + \frac{1}{2}\chi_{AB}\chi^{BC}{}_{\|C} + \frac{3}{16}(\chi_{CD}\chi^{CD})_{\|A}\right)X^A\Big]\end{aligned} \qquad \text{(C.100)}$$

(the underlined terms are again those, which potentially lead to a divergence of the integral). An integration by parts gives then Equation (6.14). Equations (C.100) and (C.99) give

$$H(X,\mathscr{S}^\epsilon) = \int_{S^2_\epsilon}\left(\mathbb{W}^{xu} - \mathbb{W}^{xA}\partial_A\alpha\right)d\theta\,d\varphi =$$

$$\frac{1}{16\pi}\int_{S^2} x^{-1}\sin\theta\,d\theta\,d\varphi\left[\chi^{AB}{}_{\|B} + \partial_u\chi^{AB}\partial_B\alpha\right]X_A + O(1)\,, \qquad \text{(C.101)}$$

(Recall that $S^2_\epsilon \equiv \partial\mathscr{S}^\epsilon$.) In this equation the derivatives $\|$ and ∂_u are derivatives with respect to the appropriate variables; however, in the above integral all the fields are taken at $u = \alpha(x=0, x^A)$, while the partial derivatives are calculated before substituting this relation, *e.g.* for a function f we have $\partial_A f(u,x^B)|_{u=\alpha} \neq \partial_A f(\alpha(x^B),x^B) = \partial_A f(u,x^B)|_{u=\alpha} + \partial_u f(\alpha,x^B)\alpha_{,A}$.

C.5 Transformation rules of the Bondi functions under supertranslations

In this appendix we wish to derive the transformation laws of the Bondi functions under supertranslations. We consider a metric of the form (5.93), and we suppose that Equations (5.98)–(5.100) hold. As elsewhere, the symbol x denotes $1/r$. From (C.4)–(C.8) and (5.98)–(5.100) one finds that the covariant (inverse) metrics has the following asymptotics

$$g(\,dx,\,dx) = x^4(1-2Mx) + O(x^6)\,, \tag{C.102}$$

$$g(\,dx,\,du) = x^2(1+\frac{1}{16}x^2\chi^{AB}\chi_{AB}) + O(x^5)\,, \tag{C.103}$$

$$g(\,dx,\,dx^A) = x^4\left[-\frac{1}{2}\chi^{AB}{}_{\|B} + x(2N^A+B^A)\right] + O(x^6)\,, \tag{C.104}$$

$$g(\,dx^B,\,dx^A) = x^2\left[\breve{h}^{AB} - x\chi^{AB} + \frac{1}{2}x^2\chi^{AC}\chi^B{}_C\right] + O(x^5)\,, \tag{C.105}$$

where

$$B^A := \frac{1}{2}\chi^A{}_B\chi^{BC}{}_{\|C} + \frac{1}{16}(\chi^{CD}\chi_{CD})^{\|A}\,. \tag{C.106}$$

We consider coordinate transformations of the form

$$\bar{u} = u - \lambda + xu_1 + x^2u_2 + x^3u_3 + \dots\,, \tag{C.107}$$

$$\bar{x} = x + x^2\rho_2 + x^3\rho_3 + \dots\,, \tag{C.108}$$

$$\bar{x}^A = x^A + xx_1^A + x^2x_2^A + x^3x_3^A + \dots\,, \tag{C.109}$$

which preserve the Bondi conditions:

$$g(\,d\bar{u},\,d\bar{u}) = g(\,d\bar{u},\,d\bar{x}^A) = 0\,, \tag{C.110}$$

$$\det\left(\bar{x}^{-2}g(\,d\bar{x}^B,\,d\bar{x}^A)\right) = \sin\bar{\theta}\,. \tag{C.111}$$

With some work one finds

$$x_1^A = \lambda^{\|A}\,, \quad u_1 = -\frac{1}{2}\lambda^{\|B}\lambda_{\|B}\,, \tag{C.112}$$

$$x_2^A = \frac{1}{2}\lambda^{\|B}\partial_B(\lambda^{\|A}) + \frac{1}{4}(\lambda^{\|B}\lambda_{\|B})^{\|A} - \frac{1}{2}\lambda_{\|B}\chi^{AB}\,, \quad \rho_2 = \frac{1}{2}\Delta_2\lambda\,, \tag{C.113}$$

$$u_2 = \frac{1}{4}\left[\chi^{AB}\lambda_{\|A}\lambda_{\|B} - \lambda^{\|A}(\lambda_{\|B}\lambda^{\|B})_{\|A}\right]\,, \tag{C.114}$$

$$\begin{aligned}\rho_3 = &-\frac{1}{2}\chi^{AB}{}_{\|B}\lambda_{\|A} - \frac{1}{4}\chi^{AB}\lambda_{\|AB} - \frac{1}{4}\dot{\chi}^{AB}\lambda_{\|A}\lambda_{\|B} + \frac{1}{4}\lambda^{\|AB}\lambda_{\|AB}\\ &+\frac{1}{2}\lambda_{\|B}\lambda^{\|B} + \frac{1}{8}(\Delta_2\lambda)^2 + \frac{1}{2}(\Delta_2\lambda)^{\|B}\lambda_{\|B}\,,\end{aligned} \tag{C.115}$$

$$\begin{aligned}x_3^A = &-\frac{1}{12}\lambda^{\|A}\dot{\chi}^{BC}\lambda_{\|B}\lambda_{\|C} - \frac{1}{12}\dot{\chi}^{AC}\lambda^{\|B}\lambda_{\|B}\lambda_{\|C}\\ &+W_3(\breve{\mathscr{D}}\lambda,\,\breve{\mathscr{D}}^2\lambda,\,\breve{\mathscr{D}}^3\lambda,\,\breve{\mathscr{D}}\chi,\chi)\,,\end{aligned} \tag{C.116}$$

where W_3 is a polynomial of degree 3. For our purposes only the time derivative $\dot{x}_3^A$ of x_3^A matters rather than x_3^A itself, equal to

$$\begin{aligned}
\dot{x}_3^A = &-\frac{1}{12}\ddot{\chi}^{CB}\lambda_{\|C}\lambda_{\|B}\lambda^{\|A} - \frac{1}{12}\ddot{\chi}^{CA}\lambda_{\|C}\lambda_{\|B}\lambda^{\|B} - \frac{1}{12}\left(\dot{\chi}^{CB}\lambda_{\|C}\lambda_{\|B}\right)^{\|A} \\
&-\frac{1}{24}\dot{\chi}_{CB}\dot{\chi}^{CB}\lambda^{\|A} - \frac{1}{6}\lambda_{\|C}\left(\lambda_{\|B}\dot{\chi}^{BA}\right)^{\|C} + \frac{1}{6}\lambda^{\|C}\left(\dot{\chi}_{CB}\chi^{BA} + \chi_{CB}\dot{\chi}^{BA}\right) \\
&-\frac{1}{3}\lambda^{\|C}\partial_B(\lambda^{\|A})\dot{\chi}_C{}^B - \frac{1}{6}\left(\lambda_{\|C}\lambda^{\|C}\right)_{\|B}\dot{\chi}^{BA} - \frac{1}{6}\lambda_{\|C}\lambda^{\|A}\dot{\chi}^{BC}{}_{\|B} \\
&-\frac{1}{12}\lambda_{\|C}\lambda^{\|C}\dot{\chi}^{BA}{}_{\|B}\,. \qquad \text{(C.117)}
\end{aligned}$$

Substituting the above in the metric, one finds the following transformation rules for the Bondi functions:

$$\bar{\chi}_{AB} = \chi_{AB} - 2\lambda_{\|AB} + \breve{h}_{AB}\Delta_2\lambda\,, \qquad \text{(C.118)}$$

$$\bar{\nabla}_{\bar{B}}\bar{\chi}^{\bar{A}\bar{B}} = \chi^{AB}{}_{\|B} - 2\dot{x}_2^A - 2\rho_2^{\|A} - 2x_1^A\,, \qquad \text{(C.119)}$$

$$\overline{M} = M + \frac{1}{2}\chi^{AB}{}_{,u\|B}\lambda_{\|A} + \frac{1}{4}\chi^{AB}{}_{,u}\lambda_{\|AB} + \frac{1}{4}\chi^{AB}{}_{,uu}\lambda_{\|A}\lambda_{\|B}\,, \qquad \text{(C.120)}$$

$$\begin{aligned}
2\bar{N}^A + \bar{B}^A = &\ 2N^A + B^A - 2Mx_1^A - 4\rho_2\rho_2{}^{\|A} + \rho_2\chi^{AB}{}_{\|B} - 2\rho_2 x_1^A - 2\rho_2\dot{x}_2^A \\
&+2x_2^A - \frac{1}{2}\chi^{CB}{}_{\|C}x_1^A{}_{,B} + \dot{\rho}_3 x_1^A + \rho_3{}^{\|A} + \dot{x}_3^A - \rho_{2,B}\chi^{AB} + \rho_2{}^{,B}x_1^A{}_{,B} \\
&+\frac{1}{2}\left[(u_1 + x_1^C\lambda_{,C})\partial_u + x_1^C\partial_C\right]\left(\chi^{AB}{}_{\|B} - 2\dot{x}_2^A - 2\rho_2^{\|A} - 2x_1^A\right)\,. \qquad \text{(C.121)}
\end{aligned}$$

In Equation (C.119) we have used a "barred" covariant derivative to emphasize the transformation law

$$\bar{\nabla}_{\bar{A}} = \nabla_A + \lambda_{,A}\partial_u\,. \qquad \text{(C.122)}$$

Using the definition (C.106) together with Equations (C.118)-(C.119) we obtain the following transformation law for B^A:

$$\begin{aligned}
\bar{B}^A = &\ B^A + \left(\alpha^{\|A}{}_B - \frac{1}{2}\delta^A{}_B\Delta_2\alpha\right)\left[(\Delta_2 + 2)\alpha\right]^{\|B} \\
&+\frac{1}{4}\left[\alpha^{\|CD}\alpha_{\|CD} - \frac{1}{2}(\Delta_2\alpha)^2\right]^{\|A} \\
&+\frac{1}{8}\alpha^{\|A}\dot{\chi}^{CD}\chi_{CD} - \frac{1}{4}\alpha^{\|A}\dot{\chi}^{CD}\alpha_{\|CD} - \frac{1}{4}\left(\chi^{CD}\alpha_{\|CD}\right)^{\|A} \\
&+\frac{1}{2}\alpha_{\|C}\dot{\chi}^{BC}\left(\chi^A{}_B - 2\alpha^{\|A}{}_B + \delta^A{}_B\Delta_2\alpha\right) - \frac{1}{2}\chi^A{}_B\left[(\Delta_2 + 2)\alpha\right]^{\|B} \\
&-\chi^{BC}{}_{\|C}\left(\alpha^{\|A}{}_B - \frac{1}{2}\delta^A{}_B\Delta_2\alpha\right)\,. \qquad \text{(C.123)}
\end{aligned}$$

The desired transformation law for N^A follows now from the equations above:

$$\begin{aligned}
2\bar{N}^A &= 2N^A - 2M\alpha^{||A} + \frac{1}{4}\left(\chi^{BC}\alpha_{||BC}\right)^{||A} + \frac{1}{2}\alpha^{||C}\chi^{AB}{}_{||BC} - \frac{1}{2}\alpha^{||C}\chi_{CB}{}^{||BA} \\
&+\frac{1}{6}\alpha^{||C}\dot{\chi}^{AB}\chi_{BC} - \frac{1}{12}\left(\dot{\chi}^{BC}\alpha_{||B}\alpha_{||C}\right)^{||A} - \frac{1}{4}\left(\dot{\chi}^{BC}\alpha_{||BC}\right)^{||A} \\
&+\frac{1}{3}\alpha^{||C}\dot{\chi}^{AB}{}_{||C}\alpha_{||B} + \frac{2}{3}\alpha^{||C}\dot{\chi}_{CB}\alpha^{||AB} - \frac{2}{3}\alpha^{||A}\alpha_{||C}\dot{\chi}^{BC}{}_{||B} \\
&+\frac{1}{6}\alpha^{||C}\alpha_{||C}\dot{\chi}^{AB}{}_{||B} - \frac{1}{6}\alpha^{||A}\dot{\chi}^{CD}\chi_{CD} - \frac{1}{3}\alpha_{||C}\dot{\chi}^{BC}\chi^{A}{}_{B} \\
&+\frac{1}{6}\alpha^{||C}\alpha_{||C}\ddot{\chi}^{AB}\alpha_{||B} - \frac{1}{3}\alpha^{||A}\alpha_{||C}\ddot{\chi}^{BC}\alpha_{||B} - \frac{1}{4}\left(\ddot{\chi}^{BC}\alpha_{||B}\alpha_{||C}\right)^{||A} .
\end{aligned} \tag{C.124}$$

C.6 Transformation rules of the Bondi functions under boosts

In this appendix we give a summary of the transformation rules of the Bondi functions under the following transformations of the "abstract Scri"

$$u \to \bar{u} = Ku , \tag{C.125}$$

with K given by the equation

$$K := \frac{\sqrt{1-\|v\|^2}}{1-v_k n^k} = \frac{1+v_k\bar{n}^k}{\sqrt{1-\|v\|^2}} , \tag{C.126}$$

as in Equation (6.45). The notation of Section 6.3 is used. These transformations correspond to asymptotic rotations or boosts of the physical space-time; we will refer collectively to those transformations as "boost transformations".

Under (C.125) the metric $\overline{h}$ corresponding to the barred Bondi angular coordinates is conformally related to the metric h as in (6.44),

$$\overline{h} = K^2 h .$$

Using the symbols Δ_2 and $\overline{\Delta}_2$ to denote the Laplace operators of h and of $\overline{h}$, one easily checks the following identities for any smooth function f:

$$\overline{\Delta}_2 f = \frac{\Delta_2 f}{K^2} , \qquad \overline{\Delta}_2(\overline{\Delta}_2+2)(Kf) = \frac{\Delta_2(\Delta_2+2)f}{K^3} .$$

The covariant derivatives of $\overline{h}$ and h will be denoted respectively by $\overline{\nabla}$ and by ∇:

$$\nabla h = \overline{\nabla}\,\overline{h} = 0 .$$

The respective Christoffel symbols are related to each other as follows:

$$\overline{\Gamma}^A_{BC} = \Gamma^A_{BC} + \delta^A_B(\ln K)_{,C} + \delta^A_C(\ln K)_{,B} - h_{BC}(\ln K)_{,D}h^{AD} .$$

With some work one finds the following transformation rules for the objects of interest:

- The shear χ_{AB} transforms as

$$\overline{\chi}_{AB}(\overline{u},\theta,\varphi) = \left(\frac{\chi_{AB}}{K^3}\right)(K(\theta,\varphi)u,\theta,\varphi)\ .$$

The arguments $(\overline{u},\theta,\varphi)$, *etc.*, in all other equations should be understood in the same way as above, and will not be repeated in what follows. The last equation leads to

$$\overline{\nabla}_A \overline{\chi}^A{}_B = \frac{1}{K^2}\nabla_A (K\chi^A{}_B)\ , \qquad \overline{\nabla}_A\overline{\nabla}_B\overline{\chi}^{AB} = \frac{1}{K^3}\nabla_A\nabla_B\chi^{AB}\ ,$$

$$\overline{\chi}^A{}_B = \frac{\chi^A{}_B}{K}\ , \qquad \overline{\chi}^{AB} = K\chi^{AB}\ , \qquad \partial_{\overline{u}}\overline{\chi}^{AB} = \partial_u\chi^{AB}\ .$$

- The mass aspect transforms as

$$\overline{M} = \frac{M}{K^3}\ .$$

A useful formula that follows from the above is the transformation law for the expression that appears in the formula for the Trautman-Bondi mass:

$$(4\overline{M} - \overline{\nabla}_A\overline{\nabla}_B\overline{\chi}^{AB})\overline{\lambda} = \frac{(4M - \nabla_A\nabla_B\chi^{AB})}{K}\lambda\ , \tag{C.127}$$

where

$$\lambda = \sqrt{\det h_{AB}}\ .$$

- Finally, the covector density λN_A is invariant under rescaling of h by K^2;

$$\overline{\lambda}\,\overline{N}_A = \lambda N_A\ . \tag{C.128}$$

C.7 Bondi coordinates in the Kerr space-time

In this appendix we will construct an approximate Bondi coordinate system for the Kerr metric. This will allow us determine the Bondi functions, and to evaluate the proposed global charges. We start with the Boyer-Lindquist coordinates (t,r,θ,φ) in which the metric functions $g_{\mu\nu}$ take the form [100, p. 878]

$$g_{tt} = -1 + \frac{2mr}{\rho^2}\ , \quad g_{t\varphi} = -\frac{2mra\sin^2\theta}{\rho^2}\ , \quad g_{rr} = \frac{\rho^2}{\triangle}\ , \quad g_{\theta\theta} = \rho^2\ ,$$

$$g_{\varphi\varphi} = \sin^2\theta\left(r^2 + a^2 + \frac{2mra^2\sin^2\theta}{\rho^2}\right)\ ,$$

where

$$\rho^2 = r^2 + a^2\cos^2\theta\ , \quad \triangle = r^2 - 2mr + a^2\ .$$

The next step is to pass to generalized *outgoing* Eddington-Finkelstein coordinates $(\tilde{u}, r, \theta, \tilde{\varphi})$ (compare [100, p. 879]), implicitly defined through the equations

$$d\tilde{u} = dt - \frac{r^2 + a^2}{\triangle}\, dr\,, \quad d\tilde{\varphi} = d\varphi - \frac{a}{\triangle}\, dr\,. \tag{C.129}$$

Equation (C.129) leads to the following non-vanishing coefficients of the metric

$$g_{\tilde{u}\tilde{u}} = -1 + \frac{2mr}{\rho^2}\,, \quad g_{\tilde{u}\tilde{\varphi}} = -\frac{2mra\sin^2\theta}{\rho^2}\,, \quad g_{r\tilde{\varphi}} = a\sin^2\theta\,, \quad g_{\theta\theta} = \rho^2\,,$$

$$g_{\tilde{\varphi}\tilde{\varphi}} = \sin^2\theta\frac{\Sigma^2}{\rho^2} = \sin^2\theta\left(r^2 + a^2 + \frac{2mra^2\sin^2\theta}{\rho^2}\right)\,, \quad g_{r\tilde{u}} = -1\,,$$

$$\Sigma^2 = (r^2 + a^2)^2 - a^2\triangle\sin^2\theta\,, \quad \sqrt{-g} = \rho^2\sin\theta\,.$$

The non-vanishing coefficients of the inverse metric read

$$g^{\tilde{u}\tilde{u}} = \frac{a^2\sin^2\theta}{\rho^2}\,, \quad g^{\tilde{u}r} = -\frac{r^2 + a^2}{\rho^2}\,, \quad g^{\tilde{u}\tilde{\varphi}} = \frac{a}{\rho^2}\,, \quad g^{rr} = \frac{\triangle}{\rho^2}\,, \quad g^{r\tilde{\varphi}} = -\frac{a}{\rho^2}\,,$$

$$g^{\theta\theta} = \frac{1}{\rho^2}\,, \quad g^{\tilde{\varphi}\tilde{\varphi}} = \frac{1}{\rho^2\sin^2\theta}\,, \quad \lambda = \sqrt{\det g_{AB}} = \Sigma\sin\theta\,,$$

and we have assumed that Σ is positive since we are only interested in the asymptotic region $r \to \infty$. Introducing$\tilde{x} := \frac{1}{r}$ we obtain:

$$g^{\tilde{u}\tilde{u}} = \frac{a^2\tilde{x}^2\sin^2\theta}{1 + a^2\tilde{x}^2\cos^2\theta}\,, \quad g^{\tilde{u}\tilde{x}} = \tilde{x}^2\frac{1 + a^2\tilde{x}^2}{1 + a^2\tilde{x}^2\cos^2\theta}\,, \quad g^{\tilde{u}\tilde{\varphi}} = \frac{a\tilde{x}^2}{1 + a^2\tilde{x}^2\cos^2\theta}\,,$$

$$g^{\tilde{x}\tilde{x}} = \tilde{x}^4\frac{1 - 2m\tilde{x} + a^2\tilde{x}^2}{1 + a^2\tilde{x}^2\cos^2\theta}\,, \quad g^{\tilde{x}\tilde{\varphi}} = \frac{a\tilde{x}^4}{1 + a^2\tilde{x}^2\cos^2\theta}\,,$$

$$g^{\theta\theta} = \frac{\tilde{x}^2}{1 + a^2\tilde{x}^2\cos^2\theta}\,, \quad g^{\tilde{\varphi}\tilde{\varphi}} = \frac{1}{\sin^2\theta}\frac{\tilde{x}^2}{1 + a^2\tilde{x}^2\cos^2\theta}\,.$$

A coordinate system asymptotic to a Bondi one is obtained by the following modification of $\tilde{u}$ and $\tilde{\varphi}$:

$$\bar{u} = \tilde{u} - \frac{1}{2}a^2\tilde{x}\sin^2\theta\,, \quad \bar{\varphi} = \tilde{\varphi} - a\tilde{x} \bmod 2\pi\,.$$

It should be stressed that the new angular coordinate $\bar{\varphi}$ is obtained from $\tilde{\varphi}$ by performing an $\tilde{x}$-dependent rotation, which corresponds to a smooth map of space-time into itself. One finds

$$g^{\bar{u}\bar{u}} = O(\tilde{x}^4)\,, \quad \tilde{x}^{-1}g^{\bar{u}\bar{\varphi}} = O(\tilde{x}^3)\,, \quad \tilde{x}^{-2}g^{\bar{u}\tilde{x}} = 1 + O(\tilde{x}^2)\,,$$

$$\tilde{x}^{-3}g^{\tilde{x}\bar{\varphi}} = 2ma\tilde{x}^2 + O(\tilde{x}^3)\,, \quad \tilde{x}^{-4}g^{\tilde{x}\tilde{x}} = \left[1 - 2m\tilde{x} + O(\tilde{x}^2)\right]\,,$$

$$\tilde{x}^{-2}g^{\theta\theta} = \left[1 + O(\tilde{x}^2)\right] , \quad \tilde{x}^{-2}g^{\bar{\varphi}\bar{\varphi}} = \frac{1}{\sin^2\theta}\left[1 + O(\tilde{x}^2)\right] .$$

It is not too difficult to check that the Bondi coordinate u asymptotic to $\bar{u}$ will satisfy

$$u = \bar{u} + O(\tilde{x}^4) ,$$

with the error term behaving in the expected way under diffentiation. Similarly the associated Bondi angular coordinate φ will be of the form

$$\varphi = \bar{\varphi} + O(\tilde{x}^4) ,$$

while the variable x related to the Bondi radial coordinate r_{Bondi} by $x := 1/r_{\mathrm{Bondi}}$ will satisfy

$$x = \tilde{x} + O(\tilde{x}^3) .$$

A straightforward calculation then shows that the associated coordinate transformation will not change the behaviour of the metric coefficients in the order that determines χ_{AB}, M and N^A. We thus obtain

$$\chi_{AB} = 0 = N^\theta , \quad M = m , \quad N^{\bar{\varphi}} = ma . \tag{C.130}$$

While this is not strictly necessary, we note the following coordinate system $(\bar{u}, \bar{x}, \bar{\theta}, \bar{\varphi})$, which approaches a Bondi one faster than the previous one:

$$\begin{aligned}
\bar{u} &= \tilde{u} - \frac{1}{2}\tilde{x}a^2\sin^2\theta + \frac{1}{8}\tilde{x}^3a^4\sin^4\theta , \\
\bar{\varphi} &= \tilde{\varphi} - \tilde{x}a + \frac{1}{3}\tilde{x}^3a^3 \bmod 2\pi , \\
\sin\bar{\theta} &= \sin\theta\left(1 + \frac{1}{2}\tilde{x}^2a^2\cos^2\theta\right) , \\
\bar{x}^2 &= \tilde{x}^2\left(1 - \tilde{x}^2a^2\sin^2\theta\right) .
\end{aligned} \tag{C.131}$$

In the coordinates (C.131) we have:

- The coordinate $\bar{u}$ approaches a lightlike one to high order in $\tilde{x}$:

$$g^{\bar{u}\bar{u}} = O(\bar{x}^5) = g^{\bar{u}\bar{A}} .$$

- The function β can be determined to order $O(\tilde{x}^3)$:

$$g^{\bar{u}\bar{x}} = \bar{x}^2 + O(\bar{x}^5) .$$

- We have

$$g^{\bar{x}\bar{\varphi}} = 2ma\bar{x}^5 + O(\bar{x}^6) , \quad g^{\bar{x}\bar{\theta}} = O(\bar{x}^6) ,$$

from which we can read U^A to high order, recovering $N^A = ma\delta^A_\varphi$.

- From
$$g^{\hat{A}\hat{B}} = \bar{x}^2 h^{\hat{A}\hat{B}} + O(\bar{x}^5)$$
we have of course again $\chi_{AB} = 0$, while the term $O(\bar{x}^5)$ suggests that the function C_{AB} from 8.4 [83, Equation 8.4, p.719] does not vanish.
- Finally
$$g^{\bar{x}\bar{x}} = \bar{x}^4 \left[1 - 2m\bar{x} + O(\bar{x}^2)\right] ,$$
$$g_{\bar{u}\bar{u}} = -1 + 2m\bar{x} + O(\bar{x}^3) ,$$
from which an approximation to V can be obtained, in particular we have $M = m$.

C.8 Conformal rescalings of ADM Cauchy data

Consider a time oriented (four-dimensional) space-time (M, g), and let $\mathscr{S} = i(\Sigma)$ be a spacelike hypersurface in M with induced metric $\gamma = i^* g$. In this Appendix we recall the transformation formulae for the extrinsic curvature K_{ij} of $i(\Sigma)$ under conformal rescalings. We set

$$L^{ij} \equiv \gamma^{ik}\gamma^{j\ell}K_{k\ell} - \frac{K}{3}\gamma^{ij} , \qquad K \equiv \gamma^{k\ell}K_{k\ell} . \tag{C.132}$$

Suppose that the metric g on M is conformally deformed by a factor Ω, $\tilde{g}_{\mu\nu} = \Omega^{-2} g_{\mu\nu}$; we set

$$\omega \equiv \Omega \circ i , \qquad \dot{\omega} \equiv n^\alpha \Omega_\alpha \circ i , \tag{C.133}$$

where n^α denotes the unit future pointing normal to $i(\Sigma)$ in (M, g). Let $\tilde{\gamma}$ denote the metric induced on Σ by $\tilde{g}$, let $\tilde{L}_{ij}$ denote (the pull-back to Σ of) the trace-free part of the extrinsic curvature $\tilde{K}_{ij}$ of $i(\Sigma)$ in $(M, \tilde{g})$, and let, finally, K denote the trace of $\tilde{K}_{ij}$:

$$\tilde{L}^{ij} \equiv \tilde{\gamma}^{ik}\tilde{\gamma}^{j\ell}\tilde{K}_{k\ell} - \frac{K}{3}\tilde{\gamma}^{ij} , \qquad K \equiv \tilde{\gamma}^{k\ell}K_{k\ell} . \tag{C.134}$$

A straightforward calculation gives

$$\tilde{L}^{ij} = \omega^3 L^{ij} , \qquad K = \omega K - 3\dot{\omega} . \tag{C.135}$$

If we set, analogously to (5.31),

$$\tilde{\pi}^{kl} := \sqrt{\det \tilde{\gamma}_{mn}}\, (\tilde{\gamma}^{ij} K_{ij} \tilde{\gamma}^{kl} - K^{kl}) , \tag{C.136}$$

we obtain

$$\tilde{\pi}^{kl} = \pi^{kl} - 2\sqrt{\det \gamma_{mn}}\gamma^{kl}\frac{\dot{\omega}}{\omega} . \tag{C.137}$$

References

1. L.F. Abbott and S. Deser, *Stability of gravity with a cosmological constant*, Nucl. Phys. **B195** (1982), 76–96.
2. R. Abraham and J.E. Marsden, *Foundations of mechanics*, Reading, Massachusetts, 1978.
3. L. Andersson, *Momenta and reduction in general relativity*, Jour. Geom. Phys. **4** (1987), 289–314.
4. L. Andersson and P.T. Chruściel, *On "hyperboloidal" Cauchy data for vacuum Einstein equations and obstructions to smoothness of null infinity*, Phys. Rev. Lett. **70** (1993), 2829–2832.
5. ———, *On "hyperboloidal" Cauchy data for vacuum Einstein equations and obstructions to smoothness of Scri*, Commun. Math. Phys. **161** (1994), 533–568.
6. ———, *On asymptotic behaviour of solutions of the constraint equations in general relativity with "hyperboloidal boundary conditions"*, Dissert. Math. **355** (1996), 1–100.
7. R. Arnowitt, S. Deser, and C. Misner, *Coordinate invariance and energy expressions in general relativity*, Phys. Rev. **122** (1961), 997–1006.
8. R. Arnowitt, S. Deser, and C.W. Misner, *The dynamics of general relativity*, Gravitation (L. Witten, ed.), Wiley, N.Y., 1962, pp. 227–265.
9. A. Ashtekar, L. Bombelli, and O. Reula, *The covariant phase space of asymptotically flat gravitational fields*, Mechanics, Analysis and Geometry: 200 years after Lagrange (M. Francaviglia, ed.), vol. 376, Elsevier Science Publishers, Amsterdam, 1991, pp. 417–450.
10. A. Ashtekar and S. Das, *Asymptotically anti-de Sitter space-times: Conserved quantities*, Class. Quantum Grav. **17** (2000), L17–L30, hep-th/9911230.
11. A. Ashtekar and A. Magnon, *Asymptotically anti–de Sitter space-times*, Classical Quantum Gravity **1** (1984), L39–L44.
12. A. Ashtekar and A. Magnon-Ashtekar, *On the symplectic structure of general relativity*, Commun. Math. Phys. **86** (1982), 55–68.
13. A. Ashtekar and M. Streubel, *Symplectic geometry of radiative modes and conserved quantities at null infinity*, Proc. Roy. Soc. London A **376** (1981), 585–607.
14. V. Balasubramanian and P. Kraus, *A stress tensor for anti-de Sitter gravity*, Commun. Math. Phys. **208** (1999), 413–428.
15. R. Bartnik, *The existence of maximal hypersurfaces in asymptotically flat space-times*, Comm. Math. Phys. **94** (1984), 155–175.
16. ———, *The mass of an asymptotically flat manifold*, Comm. Pure Appl. Math. **39** (1986), 661–693.
17. R. Bartnik, *Phase space for the Einstein equations*, unpublished (1996), URL `http://beth.canberra.edu.au/mathstat/StaffPages/Robert2.htm`.

18. R. Bartnik, P.T. Chruściel, and N. Ó Murchadha, *On maximal surfaces in asymptotically flat space-times.*, Commun. Math. Phys. **130** (1990), 95–109.
19. R. Beig, *The classical theory of canonical general relativity*, Canonical gravity: from classical to quantum (Bad Honnef, 1993) (J. Ehlers and H. Friedrich, eds.), Springer, Berlin, 1994, pp. 59–80.
20. R. Beig and P.T. Chruściel, *Killing vectors in asymptotically flat space–times: I. Asymptotically translational Killing vectors and the rigid positive energy theorem*, Jour. Math. Phys. **37** (1996), 1939–1961, gr-qc/9510015.
21. R. Beig and N. Ó Murchadha, *The Poincaré group as the symmetry group of canonical general relativity*, Ann. Phys. **174** (1987), 463–498.
22. R. Beig and W. Simon, *Proof of a multipole conjecture due to Geroch*, Commun. Math. Phys. **78** (1980), 75–82.
23. M. Berger, *Géométrie*, Nathan, Paris, 1990.
24. J. Bičák, *Selected topics in the problem of energy and radiation*, Relativity and gravitation (C.G. Kuper and A. Peres, eds.), Gordon and Breach, New York, London and Paris, 1971, N. Rosen Festschrift, pp. 47–67.
25. H. Bondi, M.G.J. van der Burg, and A.W.K. Metzner, *Gravitational waves in general relativity VII: Waves from axi–symmetric isolated systems*, Proc. Roy. Soc. London A **269** (1962), 21–52.
26. A. Borowiec, M. Ferraris, M. Francaviglia, and I. Volovich, *Energy-momentum complex for nonlinear gravitational lagrangians in the 1st-order formalism*, Gen. Rel. Grav. **26** (1994), 637–645.
27. W. Boucher, G.W. Gibbons, and G.T. Horowitz, *Uniqueness theorem for anti–de Sitter spacetime*, Phys. Rev. D **30** (1984), 2447–2451.
28. D. Brill, J. Louko, and P. Peldan, *Thermodynamics of (3+1)-dimensional black holes with toroidal or higher genus horizons*, Phys. Rev. **D56** (1997), 3600–3610, gr-qc/9705012.
29. B.D. Bramson, *Relativistic angular momentum for asymptotically flat Einstein-Maxwell manifolds*, Proc. Roy. Soc. London **A 341** (1975), 463–490.
30. J.D. Brown, S.R. Lau, and J.W. York, Jr., *Energy of isolated systems at retarded times as the null limit of quasilocal energy*, Phys. Rev. **D55** (1997), 1977–1984.
31. C. Cadeau and E. Woolgar, *New five dimensional black holes classified by horizon geometry, and a Bianchi VI braneworld*, Class. Quantum Grav. (2001), 527–542, gr-qc/0011029.
32. P.R. Chernoff and J.E. Marsden, *Properties of infinite dimensional hamiltonian systems*, Lecture Notes in Mathematics, vol. 425, Springer, New York, Heidelberg, Berlin, 1974.
33. Y. Choquet-Bruhat and R. Geroch, *Global aspects of the Cauchy problem*, Commun. Math. Phys. **14** (1969), 329–335.
34. Y. Choquet-Bruhat, J. Isenberg, and V. Moncrief, *Solutions of constraints for Einstein equations*, C. R. Acad. Sci. Paris Sér. I Math. **315** (1992), 349–355.
35. Y. Choquet-Bruhat and J. York, *The Cauchy problem*, General Relativity (A. Held, ed.), Plenum Press, New York, 1980, pp. 99–172.
36. D. Christodoulou and S. Klainerman, *The Global Nonlinear Stability of the Minkowski Space*, Princeton University Press, New Jersey, 1993.
37. D. Christodoulou and N. Ó Murchadha, *The boost problem in general relativity*, Comm. Math. Phys. **80** (1980), 271–300.
38. P.T. Chruściel, *On the relation between the Einstein and the Komar expressions for the energy of the gravitational field*, Ann. Inst. H. Poincaré **42** (1985), 267–282.
39. ———, *On angular momentum at spatial infinity*, Class. Quantum Grav. **4** (1987), L205-L210.

40. ______, *Boundary conditions at spatial infinity from a hamiltonian point of view*, Topological Properties and Global Structure of Space–Time (P. Bergmann and V. de Sabbata, eds.), Plenum Press, New York, 1986, URL `http://www.phys.univ-tours.fr/~piotr/scans`, pp. 49–59.
41. ______, *On the invariant mass conjecture in general relativity*, Commun. Math. Phys. **120** (1988), 233–248.
42. P.T. Chruściel, J. Jezierski, and M. MacCallum, *Uniqueness of scalar field energy and gravitational energy in the radiating regime*, Phys. Rev. Lett. **80** (1998), 5052–5055, gr-qc/9801073.
43. ______, *Uniqueness of the Trautman–Bondi mass*, Phys. Rev. D **58** (1998), 084001 (16 pp.), gr-qc/9803010.
44. P.T. Chruściel, M.A.H. MacCallum, and D. Singleton, *Gravitational waves in general relativity. XIV: Bondi expansions and the "polyhomogeneity" of Scri*, Phil. Trans. Roy. Soc. London A **350** (1995), 113–141.
45. P.T. Chruściel and G. Nagy, *The mass of spacelike hypersurfaces in asymptotically anti-de Sitter space-times*, (2001), in preparation.
46. P.T. Chruściel and W. Simon, *Towards the classification of static vacuum spacetimes with negative cosmological constant*, Jour. Math. Phys. **42** (2001), 1779–1817, in press, gr-qc/0004032.
47. M.A. Clayton, *Canonical general relativity: the diffeomorphism constraints and spatial frame transformations*, Jour. Math. Phys. **39** (1998), 3805–3816.
48. J. Corvino, *Scalar curvature deformation and a gluing construction for the Einstein constraint equations*, Commun. Math. Phys. **214** (2000), 137–189.
49. J. Corvino and R. Schoen, *Vacuum spacetimes which are identically Schwarzschild near spatial infinity*, talk given at the Santa Barbara Conference on Strong Gravitational Fields, June 22-26, 1999, `http://doug-pc.itp.ucsb.edu/online/gravity_c99/schoen/`.
50. S. Dain, *Initial data for stationary space-times near space-like infinity*, (2001), gr-qc/0107018.
51. S. Dain and O. Moreschi, *General existence proof for rest frame systems in asymptotically flat spacetime.*, Classical Quantum Gravity **17** (2000), 3663–3672.
52. T. Damour and B. Schmidt, *Reliability of perturbation theory in general relativity*, Jour. Math. Phys. **31** (1990), 2441–2453.
53. J. Dieudonné, *Éléments d'analyse. Tome III*, Gauthier-Villars, Paris, 1970.
54. P.A.M. Dirac, *The theory of gravitation in Hamiltonian form*, Proc. Roy. Soc. London **A246** (1958), 333–343.
55. T. Dray and M. Streubel, *Angular momentum at null infinity*, Class. Quantum Grav. **1** (1984), 15–26.
56. A. Einstein, *Das hamiltonisches Prinzip und allgemeine Relativitätstheorie*, Sitzungsber. preuss. Akad. Wiss. (1916), 1111–1116.
57. ______, *Die Grundlagen der allgemeinen Relativitätstheorie*, Ann. Phys. **49** (1916), 769–822.
58. Ph. Freud, *Über die Ausdrücke der Gesamtenergie und des Gesamtimpulses eines materiellen Systems in der allgemeinen Relativitätstheorie*, Ann. of Math., II. Ser. **40** (1939), 417–419.
59. H. Friedrich, *On the regular and the asymptotic characteristic initial value problem for Einstein's vacuum field equations*, Proc. Roy. Soc. London A **375** (1981), 169–184.
60. ______, *The asymptotic characteristic initial value problem for Einstein's vacuum field equations as an initial value problem for a first-order quasilinear symmetric hyperbolic system*, Proc. Roy. Soc. London A **378** (1981), 401–421.

61. ———, *Cauchy problem for the conformal vacuum field equations in general relativity*, Commun. Math. Phys. **91** (1983), 445–472.
62. ———, *Einstein equations and conformal structure: Existence of anti-de-Sitter-type space-times*, Jour. Geom. and Phys. **17** (1995), 125–184.
63. ———, *Hyperbolic reductions for Einstein's equations*, Class. and Quantum Grav. **13** (1996), 1451–1469.
64. ———, *Gravitational fields near space-like and null infinity*, Jour. Geom. Phys. **24** (1998), 83–163.
65. H. Friedrich and G. Nagy, *The initial boundary value problem for Einstein's vacuum field equation*, Commun. Math. Phys. **201** (1998), 619–655.
66. H. Friedrich and B.G. Schmidt, *Conformal geodesics in general relativity*, Proc. Roy. Soc. London Ser. A **414** (1987), 171–195.
67. R. Geroch, *Asymptotic structure of space-time*, Asymptotic Structure of Space-Time (F.P. Esposito and L. Witten, eds.), Plenum Press, New York, 1977, pp. 1–106.
68. G.W. Gibbons, *Gravitational entropy and the inverse mean curvature flow*, Class. Quantum Grav. **16** (1999), 1677–1687.
69. G.W. Gibbons, S.W. Hawking, G.T. Horowitz, and M.J. Perry, *Positive mass theorems for black holes*, Commun. Math. Phys. **88** (1983), 295–308.
70. D. Giulini, *Ashtekar variables in classical general relativity*, Canonical gravity: from classical to quantum (Bad Honnef, 1993) (J. Ehlers and H. Friedrich, eds.), Springer, Berlin, 1994, pp. 81–112.
71. J.D.E. Grant, *A spinorial Hamiltonian approach to gravity*, Class. Quantum Grav. **16** (1999), 3419–3437.
72. P. Hajicek and J. Kijowski, *Covariant gauge fixing and Kuchař decomposition*, Phys. Rev. **D61** (2000), 024037 (13 pp.), gr-qc/9908051.
73. R.D. Hecht and J.M. Nester, *An evaluation of the mass and spin at null infinity for the PGT and GR gravity theories*, Phys. Lett. A **217** (1996), 81–89.
74. A.D. Helfer, *A phase space for gravitational radiation*, Commun. Math. Phys. **170** (1995), 483–502.
75. M. Henneaux and C. Teitelboim, *Asymptotically anti–de Sitter spaces*, Commun. Math. Phys. **98** (1985), 391–424.
76. D. Hilbert, *Die Grundlagen der Physik*, Nachrichten von der Gesellschaft der Wissenschaften zu Göttingen. Mathematisch–physikalische Klasse (1915), 395–407.
77. ———, *Die Grundlagen der Physik*, Nachrichten von der Gesellschaft der Wissenschaften zu Göttingen. Mathematisch–physikalische Klasse (1917), 53–76.
78. G.T. Horowitz and R.C. Myers, *The AdS/CFT correspondence and a new positive energy conjecture for general relativity*, Phys. Rev. **D59** (1999), 026005 (12 pp.).
79. J. Isenberg and V. Moncrief, *Some results on non–constant mean curvature solutions of the Einstein constraint equations*, Physics on Manifolds (M. Flato, R. Kerner, and A. Lichnerowicz, eds.), Kluwer Academic Publishers, Dordrecht, 1994, Y. Choquet–Bruhat Festschrift, pp. 295–302.
80. J. Isenberg and J. Nester, *Canonical gravity*, General relativity and gravitation, Vol. 1, Plenum, New York, 1980, pp. 23–97.
81. J. Jezierski, *Positivity of Mass for Certain Spacetimes with Horizons*, Classical and Quantum Gravity **6** (1989), 1535–1539.
82. ———, *Perturbation of initial data for spherically symmetric charged black hole and Penrose conjecture*, Acta Phys. Polonica B **25** (1994), 1413–17
83. ———, *Bondi mass in classical field theory*, Acta Phys. Polonica B **29** (1998), 667–743, gr-qc/9703083.

84. ______, *Energy and angular momentum of the weak gravitational waves on the Schwarzschild background — quasi-local gauge invariant formulation*, Gen. Rel. Grav. **31** (1999), 1855–1890, gr-qc/9801068.
85. J. Jezierski and J. Kijowski, *Positivity of total energy in general relativity*, Phys. Rev. **D36** (1987), 1041–1044.
86. B. Julia and S. Silva, *Currents and superpotentials in classical gauge invariant theories. i: Local results with applications to perfect fluids and general relativity*, Class. Quant. Grav. **15** (1998), 2173–2215, gr-qc/9804029.
87. ______, *Currents and superpotentials in classical gauge theories. II: Global aspects and the example of affine gravity*, Class. Quantum Grav. **17** (2000), 4733–4744.
88. J. Kánnár, *Hyperboloidal initial data for the vacuum Einstein equations with cosmological constant*, Class. Quantum Grav. **13** (1996), 3075–3084.
89. ______, *On the existence of C^∞ solutions to the asymptotic characteristic initial value problem in general relativity*, Proc. Roy. Soc. London A **452** (1996), 945–952.
90. J. Katz, J. Bičák, and D. Lynden-Bell, *Relativistic conservation laws and integral constraints for large cosmological perturbations*, Phys. Rev. D **55** (1997), 5957–5969.
91. J. Kijowski, *On a new variational principle in general relativity and the energy of the gravitational field*, Gen. Rel. Grav. **9** (1978), 857–877.
92. ______, *Unconstrained degrees of freedom of gravitational field and the positivity of gravitational energy*, Gravitation, geometry and relativistic physics (Aussois, 1984), Springer, Berlin, 1984, pp. 40–50.
93. ______, *Asymptotic degrees of freedom and gravitational energy*, Proceedings of Journées Relativistes 1983 (Bologna) (S. Benenti et al., ed.), Pitagora Editrice, 1985, pp. 205–211.
94. ______, *A simple derivation of canonical structure and quasi-local Hamiltonians in general relativity*, Gen. Rel. Grav. **29** (1997), 307–343.
95. J. Kijowski and W.M. Tulczyjew, *A symplectic framework for field theories*, Lecture Notes in Physics, vol. 107, Springer, New York, Heidelberg, Berlin, 1979.
96. F. Kottler, *Über die physikalischen Grundlagen der Einsteinschen Gravitationstheorie*, Annalen der Physik **56** (1918), 401–462.
97. K. Kuchař, *A bubble-time canonical formalism for geometrodynamics*, Jour. Math. Phys. **13** (1972), 768–781.
98. P. Libermann and C.M. Marle, *Symplectic geometry and analytical mechanics*, Mathematics and its Applications,, vol. 35, D. Reidel Publ. Co., Dordrecht, 1987, transl. from the french by B.E. Schwarzbach.
99. A. Magnon, *On Komar integrals in asymptotically anti–de Sitter space-times*, Jour. Math. Phys. **26** (1985), 3112–3117.
100. C.W. Misner, K. Thorne, and J.A. Wheeler, *Gravitation*, Freeman, San Fransisco, 1973.
101. N. Ó Murchadha, *Total energy momentum in general relativity*, J. Math. Phys. **27** (1986), 2111–2128.
102. J. M. Nester, *The gravitational Hamiltonian*, Asymptotic behaviour of mass and space-time geometry (Oregon 1983) (F. J. Flaherty, ed.), Lecture Notes in Physics 212, Springer Verlag, 1984, pp. 155–163.
103. A. Palatini, *Deduzione invariantive delle equazioni gravitazionali dal principio di Hamilton*, Rend. Circ. Mat. Palermo **43** (1919), 203–212.
104. R. Penrose, *Asymptotic properties of fields and space–times*, Phys. Rev. Lett. **10** (1963), 66–68.

105. ———, *Zero rest-mass fields including gravitation*, Proc. Roy. Soc. London **A284** (1965), 159–203.
106. ———, *Relativistic symmetry groups*, Group theory in non-linear problems (A.O. Barut, ed.), D. Reidel, Dordrecht, 1974.
107. T. Regge and C. Teitelboim, *Role of surface integrals in the Hamiltonian formulation of general relativity*, Ann. Phys. **88** (1974), 286–318.
108. A. Rizzi, *Angular momentum in general relativity: a new definition*, Phys. Rev. Lett. **81** (1998), no. 6, 1150–1153.
109. R. Sachs, *Gravitational waves in general relativity VIII. Waves in asymptotically flat space-time*, Proc. Roy. Soc. London A **270** (1962), 103–126.
110. D.J. Saunders, *The geometry of jet bundles*, vol. (LMS 142), Cambridge Univ. Press, 1989.
111. B. Schmidt, M. Walker, and P. Sommers, *A characterization of the Bondi-Metzner-Sachs group*, Gen. Rel. Grav. **6** (1975), 489–497.
112. R. Schoen and S.-T. Yau, *Proof of the positive mass theorem II*, Comm. Math. Phys. **79** (1981), 231–260.
113. J.A. Schouten, *Tensor Analysis for Physicists*, Oxford, Clarendon Press, 1951.
114. S. Silva, *On superpotentials and charge algebras of gauge theories*, Nucl. Phys. **B558** (1999), 391–415, hep-th/9809109.
115. R.D. Sorkin, *Conserved quantities as action variations*, Mathematics and general relativity (Santa Cruz, CA, 1986), Amer. Math. Soc., Providence, RI, 1988, pp. 23–37.
116. M. Spivak, *A comprehensive introduction to differential geometry. Vol. V*, second ed., Publish or Perish Inc., Wilmington, Del., 1979.
117. L. Szabados, *On certain quasi-local spin-angular momentum expressions for small spheres*, Class. Quantum Grav. **13** (1999), 2889–2904.
118. L.A. Tamburino and J.H. Winicour, *Gravitational fields in finite and conformal Bondi frames*, Phys. Rev. **150** (1966), 1039–1053.
119. M.E. Taylor, *Partial differential equations*, Springer, New York, Berlin, Heidelberg, 1996.
120. A. Trautman, *King's College lecture notes on general relativity, May-June 1958*, mimeographed notes; to be reprinted in *Gen. Rel. Grav.*
121. ———, *Radiation and boundary conditions in the theory of gravitation*, Bull. Acad. Pol. Sci., Série sci. math., astr. et phys. **VI** (1958), 407–412.
122. P. Urbański, *Analiza dla studentów fizyki*, Uniwersytet Warszawski, 1998 (in polish).
123. M.G.J. van der Burg, *Gravitational waves in general relativity IX. Conserved quantities*, Proc. Roy. Soc. London A **294** (1966), 112–122.
124. L. Vanzo, *Black holes with unusual topology*, Phys. Rev. **D56** (1997), 6475–6483, gr-qc/9705004.
125. R.M. Wald, *General relativity*, University of Chicago Press, Chicago, 1984.
126. R.M. Wald and A. Zoupas, *A general definition of "conserved quantities" in general relativity and other theories of gravity*, Phys. Rev. **D61** (2000), 084027 (16 pp.), gr-qc/9911095.
127. E. Witten, *A simple proof of the positive energy theorem*, Comm. Math. Phys. **80** (1981), 381–402.

Lecture Notes in Physics

For information about Vols. 1–543
please contact your bookseller or Springer-Verlag

Vol. 544: T. Brandes (Ed.), Low-Dimensional Systems. Interactions and Transport Properties. Proceedings, 1999. VIII, 219 pages. 2000

Vol. 545: J. Klamut, B. W. Veal, B. M. Dabrowski, P. W. Klamut, M. Kazimierski (Eds.), New Developments in High-Temperature Superconductivity. Proceedings, 1998. VIII, 275 pages. 2000.

Vol. 546: G. Grindhammer, B. A. Kniehl, G. Kramer (Eds.), New Trends in HERA Physics 1999. Proceedings, 1999. XIV, 460 pages. 2000.

Vol. 547: D. Reguera, G. Platero, L. L. Bonilla, J. M. Rubí(Eds.), Statistical and Dynamical Aspects of Mesoscopic Systems. Proceedings, 1999. XII, 357 pages. 2000.

Vol. 548: D. Lemke, M. Stickel, K. Wilke (Eds.), ISO Surveys of a Dusty Universe. Proceedings, 1999. XIV, 432 pages. 2000.

Vol. 549: C. Egbers, G. Pfister (Eds.), Physics of Rotating Fluids. Selected Topics, 1999. XVIII, 437 pages. 2000.

Vol. 550: M. Planat (Ed.), Noise, Oscillators and Algebraic Randomness. Proceedings, 1999. VIII, 417 pages. 2000.

Vol. 551: B. Brogliato (Ed.), Impacts in Mechanical Systems. Analysis and Modelling. Lectures, 1999. IX, 273 pages. 2000.

Vol. 552: Z. Chen, R. E. Ewing, Z.-C. Shi (Eds.), Numerical Treatment of Multiphase Flows in Porous Media. Proceedings, 1999. XXI, 445 pages. 2000.

Vol. 553: J.-P. Rozelot, L. Klein, J.-C. Vial Eds.), Transport of Energy Conversion in the Heliosphere. Proceedings, 1998. IX, 214 pages. 2000.

Vol. 554: K. R. Mecke, D. Stoyan (Eds.), Statistical Physics and Spatial Statistics. The Art of Analyzing and Modeling Spatial Structures and Pattern Formation. Proceedings, 1999. XII, 415 pages. 2000.

Vol. 555: A. Maurel, P. Petitjeans (Eds.), Vortex Structure and Dynamics. Proceedings, 1999. XII, 319 pages. 2000.

Vol. 556: D. Page, J. G. Hirsch (Eds.), From the Sun to the Great Attractor. X, 330 pages. 2000.

Vol. 557: J. A. Freund, T. Pöschel (Eds.), Stochastic Processes in Physics, Chemistry, and Biology. X, 330 pages. 2000.

Vol. 558: P. Breitenlohner, D. Maison (Eds.), Quantum Field Theory. Proceedings, 1998. VIII, 323 pages. 2000

Vol. 559: H.-P. Breuer, F. Petruccione (Eds.), Relativistic Quantum Measurement and Decoherence. Proceedings, 1999. X, 140 pages. 2000.

Vol. 560: S. Abe, Y. Okamoto (Eds.), Nonextensive Statistical Mechanics and Its Applications. IX, 272 pages. 2001.

Vol. 561: H. J. Carmichael, R. J. Glauber, M. O. Scully (Eds.), Directions in Quantum Optics. XVII, 369 pages. 2001.

Vol. 562: C. Lämmerzahl, C. W. F. Everitt, F. W. Hehl (Eds.), Gyros, Clocks, Interferometers...: Testing Relativistic Gravity in Space. XVII,507 pages. 2001.

Vol. 563: F. C. Lázaro, M. J. Arévalo (Eds.), Binary Stars. Selected Topics on Observations and Physical Processes. 1999.IX, 327 pages. 2001.

Vol. 564: T. Pöschel, S. Luding (Eds.), Granular Gases. VIII, 457 pages. 2001.

Vol. 565: E. Beaurepaire, F. Scheurer, G. Krill, J.-P. Kappler (Eds.), Magnetism and Synchrotron Radiation. XIV, 388 pages. 2001.

Vol. 566: J. L. Lumley (Ed.), Fluid Mechanics and the Environment: Dynamical Approaches. VIII, 412 pages. 2001.

Vol. 567: D. Reguera, L. L. Bonilla, J. M. Rubí (Eds.), Coherent Structures in Complex Systems. IX, 465 pages. 2001.

Vol. 568: P. A. Vermeer, S. Diebels, W. Ehlers, H. J. Herrmann, S. Luding, E. Ramm (Eds.), Continuous and Discontinuous Modelling of Cohesive-Frictional Materials. XIV, 307 pages. 2001.

Vol. 569: M. Ziese, M. J. Thornton (Eds.), Spin Electronics. XVII, 493 pages. 2001.

Vol. 570: S. G. Karshenboim, F. S. Pavone, F. Bassani, M. Inguscio, T. W. Hänsch (Eds.), The Hydrogen Atom: Precision Physics of Simple Atomic Systems. XXIII, 293 pages. 2001.

Vol. 571: C. F. Barenghi, R. J. Donnelly, W. F. Vinen (Eds.), Quantized Vortex Dynamics and Superfluid Turbulence. XXII, 455 pages. 2001.

Vol. 572: H. Latal, W. Schweiger (Eds.), Methods of Quantization. XI, 224 pages. 2001.

Vol. 573: H. M. J. Boffin, D. Steeghs, J. Cuypers (Eds.), Astrotomography. XX, 434 pages. 2001.

Vol. 574: J. Bricmont, D. Dürr, M. C. Galavotti, G. Ghirardi, F. Petruccione, N. Zanghi (Eds.), Chance in Physics. XI, 288 pages. 2001.

Vol. 575: M. Orszag, J. C. Retamal (Eds.), Modern Challenges in Quantum Optics. XXIII, 405 pages. 2001.

Vol. 576: M. Lemoine, G. Sigl (Eds.), Physics and Astrophysics of Ultra High Energy Cosmic Rays. X, 318 pages. 2001.

Vol. 577: N. Thomas, I. P. Williams (Eds.), Solar and Extra-Solar Planetary Systems. X, 266 pages. 2001.

Vol. 578: D. Blaschke, N. K. Glendenning, A. Sedrakian (Eds.), Physics of Neutron Star Interiors. XI, 509 pages. 2001.

Vol. 579: R. Haug, H. Schoeller (Eds.), Interacting Electrons in Nanostructures. X, 227 pages. 2001.

Vol. 580: K. Baberschke, M. Donath, W. Nolting (Eds.), Band-Ferromagnetism: Ground-State and Finite-Temperature Phenomena.XI, 346 pages. 2001.

Vol.582: N. J. Balmforth, A. Provenzale (Eds.), Fluid Mechanical Problems in Geomorphology. VII, 578 pages. 2001.

Monographs

For information about Vols. 1–28
please contact your bookseller or Springer-Verlag

Vol. m 29: C. Bendjaballah, Introduction to Photon Communication. VII, 193 pages. 1995.

Vol. m 30: A. J. Greer, W. J. Kossler, Low Magnetic Fields in Anisotropic Superconductors. VII, 161 pages. 1995.

Vol. m 31 (Corr. Second Printing): P. Busch, M. Grabowski, P.J. Lahti, Operational Quantum Physics. XII, 230 pages. 1997.

Vol. m 32: L. de Broglie, Diverses questions de mécanique et de thermodynamique classiques et relativistes. XII, 198 pages. 1995.

Vol. m 33: R. Alkofer, H. Reinhardt, Chiral Quark Dynamics. VIII, 115 pages. 1995.

Vol. m 34: R. Jost, Das Märchen vom Elfenbeinernen Turm. VIII, 286 pages. 1995.

Vol. m 35: E. Elizalde, Ten Physical Applications of Spectral Zeta Functions. XIV, 224 pages. 1995.

Vol. m 36: G. Dunne, Self-Dual Chern-Simons Theories. X, 217 pages. 1995.

Vol. m 37: S. Childress, A.D. Gilbert, Stretch, Twist, Fold: The Fast Dynamo. XI, 406 pages. 1995.

Vol. m 38: J. González, M. A. Martín-Delgado, G. Sierra, A. H. Vozmediano, Quantum Electron Liquids and High-Tc Superconductivity. X, 299 pages. 1995.

Vol. m 39: L. Pittner, Algebraic Foundations of Non-Com-mutative Differential Geometry and Quantum Groups. XII, 469 pages. 1996.

Vol. m 40: H.-J. Borchers, Translation Group and Particle Representations in Quantum Field Theory. VII, 131 pages. 1996.

Vol. m 41: B. K. Chakrabarti, A. Dutta, P. Sen, Quantum Ising Phases and Transitions in Transverse Ising Models. X, 204 pages. 1996.

Vol. m 42: P. Bouwknegt, J. McCarthy, K. Pilch, The W3 Algebra. Modules, Semi-infinite Cohomology and BV Algebras. XI, 204 pages. 1996.

Vol. m 43: M. Schottenloher, A Mathematical Introduction to Conformal Field Theory. VIII, 142 pages. 1997.

Vol. m 44: A. Bach, Indistinguishable Classical Particles. VIII, 157 pages. 1997.

Vol. m 45: M. Ferrari, V. T. Granik, A. Imam, J. C. Nadeau (Eds.), Advances in Doublet Mechanics. XVI, 214 pages. 1997.

Vol. m 46: M. Camenzind, Les noyaux actifs de galaxies. XVIII, 218 pages. 1997.

Vol. m 47: L. M. Zubov, Nonlinear Theory of Dislocations and Disclinations in Elastic Body. VI, 205 pages. 1997.

Vol. m 48: P. Kopietz, Bosonization of Interacting Fermions in Arbitrary Dimensions. XII, 259 pages. 1997.

Vol. m 49: M. Zak, J. B. Zbilut, R. E. Meyers, From Instability to Intelligence. Complexity and Predictability in Nonlinear Dynamics. XIV, 552 pages. 1997.

Vol. m 50: J. Ambjørn, M. Carfora, A. Marzuoli, The Geometry of Dynamical Triangulations. VI, 197 pages. 1997.

Vol. m 51: G. Landi, An Introduction to Noncommutative Spaces and Their Geometries. XI, 200 pages. 1997.

Vol. m 52: M. Hénon, Generating Families in the Restricted Three-Body Problem. XI, 278 pages. 1997.

Vol. m 53: M. Gad-el-Hak, A. Pollard, J.-P. Bonnet (Eds.), Flow Control. Fundamentals and Practices. XII, 527 pages. 1998.

Vol. m 54: Y. Suzuki, K. Varga, Stochastic Variational Approach to Quantum-Mechanical Few-Body Problems. XIV, 324 pages. 1998.

Vol. m 55: F. Busse, S. C. Müller, Evolution of Spontaneous Structures in Dissipative Continuous Systems. X, 559 pages. 1998.

Vol. m 56: R. Haussmann, Self-consistent Quantum Field Theory and Bosonization for Strongly Correlated Electron Systems. VIII, 173 pages. 1999.

Vol. m 57: G. Cicogna, G. Gaeta, Symmetry and Perturbation Theory in Nonlinear Dynamics. XI, 208 pages. 1999.

Vol. m 58: J. Daillant, A. Gibaud (Eds.), X-Ray and Neutron Reflectivity: Principles and Applications. XVIII, 331 pages. 1999.

Vol. m 59: M. Kriele, Spacetime. Foundations of General Relativity and Differential Geometry. XV, 432 pages. 1999.

Vol. m 60: J. T. Londergan, J. P. Carini, D. P. Murdock, Binding and Scattering in Two-Dimensional Systems. Applications to Quantum Wires, Waveguides and Photonic Crystals. X, 222 pages. 1999.

Vol. m 61: V. Perlick, Ray Optics, Fermat's Principle, and Applications to General Relativity. X, 220 pages. 2000.

Vol. m 62: J. Berger, J. Rubinstein, Connectivity and Superconductivity. XI, 246 pages. 2000.

Vol. m 63: R. J. Szabo, Ray Optics, Equivariant Cohomology and Localization of Path Integrals. XII, 315 pages. 2000.

Vol. m 64: I. G. Avramidi, Heat Kernel and Quantum Gravity. X, 143 pages. 2000.

Vol. m 65: M. Hénon, Generating Families in the Restricted Three-Body Problem. Quantitative Study of Bifurcations. XII, 301 pages. 2001.

Vol. m 66: F. Calogero, Classical Many-Body Problems Amenable to Exact Treatments. XIX, 749 pages. 2001.

Vol. m 67: A. S. Holevo, Statistical Structure of Quantum Theory. IX, 159 pages. 2001.

Vol. m 68: N. Polonsky, Supersymmetry: Structure and Phenomena. Extensions of the Standard Model. XV, 169 pages. 2001.

Vol. m 69: W. Staude, Laser-Strophometry. High-Resolution Techniques for Velocity Gradient Measurements in Fluid Flows. XV, 178 pages. 2001.

Vol. m 70: P. T. Chruściel, J. Jezierski, J. Kijowski, Hamiltonian Field Theory in the Radiating Regime. VI, 172 pages. 2002.